P9-DOB-657

AMERICAN SHERLOCK

Also by Kate Winkler Dawson

Death in the Air: The True Story of a Serial Killer, the Great London Smog, and the Strangling of a City

AMERICAN SHERLOCK

Murder, Forensics, and
the Birth of American CSI

KATE WINKLER DAWSON

G. P. Putnam's Sons | New York

PUTNAM
— EST. 1838 —

G. P. Putnam's Sons
Publishers Since 1838
An imprint of Penguin Random House LLC
penguinrandomhouse.com

Hardcover ISBN 9780525539551
Ebook ISBN 9780525539575

Printed in the United States of America
1 3 5 7 9 10 8 6 4 2

Book design by Laura K. Corless

To Quinn and Ella, our family's greatest storytellers

CONTENTS

PROLOGUE:
Tales from the Archive: Pistols, Jawbones, and Love Poetry
1

CHAPTER 1:
A Bloody Mess: The Case of Allene Lamson's Bath, Part I
7

CHAPTER 2:
Genius: The Case of Oscar Heinrich's Demons
25

CHAPTER 3:
Heathen: The Case of the Baker's Handwriting, Part I
43

CHAPTER 4:
Pioneer: The Case of the Baker's Handwriting, Part II
65

CHAPTER 5:
Damnation: The Case of the Star's Fingerprints, Part I
87

CHAPTER 6:
Indignation: The Case of the Star's Fingerprints, Part II
111

CHAPTER 7:
Double 13: The Case of the Great Train Heist
129

CHAPTER 8:
Bad Chemistry: The Case of the Calculating Chemist
163

CHAPTER 9:
Bits and Pieces: The Case of Bessie Ferguson's Ear
185

CHAPTER 10:
Triggered: The Case of Marty Colwell's Gun
207

CHAPTER 11:
Damned: The Case of Allene Lamson's Bath, Part II
227

EPILOGUE:
Case Closed
267

Acknowledgments
275

Notes
279

Index
315

AMERICAN SHERLOCK

PROLOGUE

Tales from the Archive: Pistols, Jawbones, and Love Poetry

His upper jawbone was massive—a long, curved bone with nine tiny holes meant to hold his teeth. The remainder of his skeleton was blackened by a fairly large fire ignited by an anonymous killer. Lifting up the jawbone, I examined the small blades of grass that adhered to its exterior—organic evidence from his hillside grave in El Cerrito in Northern California.

It was distressing to hold a bone that had belonged to a murder victim, particularly one who was never identified. I glanced over at the archivist, Lara Michels, who quietly stood across the wooden desk inside the massive warehouse. "What's next?" I asked.

She led me down a long row of large cartons, more than one hundred boxes donated by the same owner. I had been given exclusive access to a trove of material collected over five decades by a brilliant man, a forensic scientist and criminalist from the first half of the twentieth century, a man who changed how crimes were solved before forensics became the foundation of most criminal cases—America's Sherlock Holmes. I walked along the tight corridor, scanning the labels on the cardboard boxes for a common name: Edward Oscar Heinrich.

When Heinrich died in 1953, at the age of seventy-two, his youngest child, Mortimer, waited sixteen years to donate the contents of his father's laboratory, a bastion of forensic history that once monopolized the ground floor of Mortimer's childhood home in Berkeley, California. In 1968, he bequeathed his father's many boxes, containing case files, evidence, personal diaries, letters, even romantic poetry, to the University of California at Berkeley, Oscar's alma mater and the college where he spent years teaching forensic science. The archive was an incredible repository of information, but given the university's limited budget for archival material and research, the collection remained uncatalogued and untouched for more than fifty years.

In 2016, I discovered Oscar Heinrich hidden in a short article that lauded one of his most famous cases, the Siskiyou train robbery of 1923. Astonished that no contemporary author had penned a book about him, I requested that UC Berkeley open his collection for research. Michels agreed, and after more than a year of waiting, I began to immerse myself in the bizarre world of Oscar Heinrich, the most famous criminalist you've likely never heard of.

The boxes contain more than one hundred thousand pieces of information, such as photographs, notes, letters, sketches, and trial transcripts. It was an overwhelming and disorganized collection that was housed in the school's off-site processing center. Heinrich seemingly kept *everything* from his life (personal and professional), manically collecting notes written on napkins, thousands of newspapers, hundreds of bullets, and dozens of financial journals. I began jokingly describing him as a "productive hoarder"—until my colleague, a psychology professor, at the University of Texas suggested that he had in fact fit the diagnostic criteria for obsessive-compulsive personality disorder, which occurs in just 1 percent of the population. People with OCPD have a preoccupation with perfectionism, control, and order—a neat life. They are frequently extremely productive and successful, but their personal relationships often suffer because their rigidity can manifest itself in righteousness, even anger when their control is threatened. Heinrich's already-stressful life was certainly complicated by his OCPD, but as an author and researcher, I was thankful for his fastidious habit of adding constantly to his collection. I was particularly grateful for the numerous boxes of evidence he had preserved from criminal cases.

The evidence was plentiful, spanning investigations that unraveled over decades. The archivist allowed me to examine pieces from a detonated bomb, a locket owned by a dead woman who was run down by her own car, a lock of hair belonging to an actress who died during an infamous party, and several pistols that required having their firing pins removed by UC Berkeley police.

As I picked up the first photo, I was struck by something that seemed like an odd observation at the time—Heinrich was quite handsome for a tightly wound scientist. He was slight and not particularly tall, with thinning light brown hair. There was something about the sharp angles of his face that made him magnetic in photos, a confidence in his eyes as he cleaned a revolver.

I spent months staring at thousands of photographs, some taken by Heinrich's assistants and others developed by the criminalist himself (he was an avid photographer who relished documenting crime scenes). I noted hundreds of details, like the way he squinted as he adjusted the focus knob on his favorite microscope. The way his teeth gripped the bit of a straight-stemmed pipe as a small stream of smoke billowed from its bowl. The way his forehead wrinkled as he hunched over evidence. The way his round rimless glasses fit extra-snuggly around his temples—a requirement for a chemist who spent much of his time leaning over a microscope.

As I flipped through those portraits, I gleaned more details about his private lab in Berkeley Hills, a lovely neighborhood overlooking San Francisco Bay. Heinrich was surrounded by odd devices. Every conceivable type of microscope was crammed onto a long wooden desk. Any extra space was surrendered to test tubes, crucibles, beakers, lenses, and scales. Behind Heinrich were shelves filled with hundreds of priceless books, at least priceless to a chemist turned forensic scientist. There were tomes on fingerprint identification, applied mechanics, analytic geometry, and powdered vegetable drugs.

The titles, written in six different languages, would intrigue any intellectual. *Blood, Urine, Feces and Moisture: A Book of Tests,* read one cover. *Arsenic in Papers and Fabrics,* read another. He even owned a tattered dictionary of slang used by criminals. They seemed unrelated, a cache of mismatched textbooks in the library of a brilliant madman. But each was a tiny piece belonging to a bigger puzzle that only he could assemble. The portrait of a genius and the tumultuous era in which he lived began to emerge.

And it *was* a tumultuous era—the homicide rate in the 1920s, when Heinrich's most interesting work began, had increased by as much as almost 80 percent from the decade before, thanks to Prohibition. For thirteen years the federal government banned alcohol in hopes of reducing crime, but instead it spawned new and more creative criminal enterprises. Varying levels of corruption tainted local governments and police departments across the country. Judges enjoyed immunity from arrest, and most major cities were ruled by crime bosses. Poverty and unemployment were also responsible for the increase in violent crimes, as many Americans became desperate for security and safety. And there was an ever-growing backlog of unsolved crimes.

The FBI was still the Bureau of Investigation, a group of insufficiently trained officers who mostly investigated bank fraud. Local police forces were underfunded, poorly instructed, and mostly using investigative techniques that hadn't been updated since the Victorian era. There would be no public federal crime lab until 1932; violent bank robberies increased while murderers terrorized Americans, especially women, whose newfound independence inflamed both the passions and the anger of many in society.

The archaic methods of crime fighting in the 1920s, procedures depending on hunches and weak circumstantial evidence, were futile. Cops were combatting a sneakier criminal, those thieves and murderers who understood chemicals, firearms, and the criminal court system. Police were outmanned and many times outsmarted.

"Footprints are the best clue," declared one top cop at the time. "There's no need for any other type of identification."

Innocent men were being hanged while criminals escaped justice. The complicated crimes of the 1920s demanded a special type of sleuth—an expert with the instincts of a detective in the field, the analytical skills of a forensic scientist in the lab, and the ability to translate that knowledge to a general audience in a courtroom. Edward Oscar Heinrich became the nation's first unique crime scene investigator—one of America's greatest forensic scientists, a criminalist who cracked some of the country's most baffling cases.

But not everyone in law enforcement welcomed his peculiar approach. In 1910, when he opened the nation's first private crime lab in Tacoma, Washington, he was scorned and quickly labeled a quack, an arrogant academic who claimed he could solve baffling crimes with some suspicious chemicals and a heavy microscope. His snappy tweed suits made him seem more like a square college professor than a seasoned detective. But he offered astounding results, solving at least two thousand cases in his more than forty-year career. He would regularly work between thirty and forty cases a month.

The press at the time dubbed Edward Oscar Heinrich "America's Sherlock Holmes" thanks to his brilliance in the lab, his cool demeanor at crime scenes, and his expertise in the witness chair. Between 1921 and 1933, his reputation evolved from curiosity to legend. His cases are enshrined in books, but their hero is largely unknown—a pioneer in the world of crime solving whose fingerprint is everywhere.

He invented new forensic techniques, a CSI in the field and inside the lab before the acronym even existed. And he was a nascent innovator of criminal profiling fifty years before the FBI's Behavioral Science Unit invented its methodology, in 1972. Present-day scientists recite his methods as they sit on the stand in criminal cases. He pioneered countless methods that we take for granted as part of the crime-fighting arsenal—techniques like blood-spatter analysis, ballistics, and latent fingerprint retrieval and analysis. It's safe to say Oscar Heinrich shaped modern criminal investigation techniques as much as any other scientist in the twentieth century.

He also pioneered some significant mistakes—problems that law enforcement are still grappling with today.

So much can be gleaned from Heinrich's best-known cases, many of which were front-page news at the time (but most of which have fallen into obscurity, much like the man himself). It was through these cases that his reputation was made. It was also in a few key cases that his worst mistakes were codified for generations of investigators to come. But first, to understand where Heinrich went wrong, we need to understand where he went right, by peering into his work at the very height of his powers.

1.

A Bloody Mess: The Case of Allene Lamson's Bath, Part I

He dipped into this bottle or that, drawing out a few drops of each with his glass pipette, and finally brought a test-tube containing a solution over to the table. . . . "You come at a crisis, Watson," said he. "If this paper remains blue, all is well. If it turns red, it means a man's life."

—Arthur Conan Doyle,
The Naval Treaty, 1893

The sharp crackles in the back garden signaled a weekend ritual—the sporadic popping from a small fire, one of many bonfires in her yard over the past three years. Her husband was fond of burning the rubbish he collected from their small bungalow-style home in Northern California.

It was Tuesday, May 30, 1933. The fire sizzled, consuming an incredible amount of debris: garden trimmings, dead artichoke plants, long-dead snails, useless paper, pieces of canvas, and even old steak bones—anything David Lamson thought might reduce to ash by late morning. The pungent smell grew stronger, like charred meat served by a distracted chef, but Allene Lamson rarely complained. The fires helped satisfy her husband's compulsion to keep their home orderly.

It was an honor to live along Stanford University's prestigious

Faculty Row in Palo Alto, an affluent community about thirty miles south of San Francisco. Now a high-tech hub in the heart of Silicon Valley, the city has always attracted the wealthy, the educated, and the kingmakers, even in the 1930s. The Lamsons' cottage was snuggled amid the palatial homes of professors and professionals, surrounded by the splendid coast live oaks and flowering eucalyptus trees on campus. The university had earned an international reputation by the 1930s—a sanctuary for future academics who could afford a pricey private education, even as most Americans struggled through the fourth year of the Great Depression, later called the toughest year.

The Lamsons' cottage on Salvatierra Street, with its Spanish-style red-tiled roof and stucco walls adorned with ivy, was modest compared to the other lavish homes in the neighborhood. The house was just a ten-minute stroll from former president Herbert Hoover's impressive three-tiered residence. His wife, First Lady Lou Henry, had an interest in architecture; in 1919, she'd helped to design the five-thousand-square-foot home in the newly popular International style of European estates. In the 1920s, she had overseen the construction of seven single-story cottages on the Row for younger faculty, with prices ranging from about $4,000 to $7,000, and the Lamsons had purchased one.

President Hoover had recently retreated to his sprawling California estate after being soundly defeated in the last election by Democrat Franklin D. Roosevelt. Many Americans blamed Hoover for the Great Depression, the catastrophic economic collapse triggered by the stock market crash just seven months after the Republican took office in 1929. By 1933, shantytowns called "Hoovervilles" increasingly dotted America. Bread lines and soup kitchens served millions of impoverished people as Hoover returned to Palo Alto with a tainted legacy. While the former president's two-acre property might have seemed ostentatious, the Lamsons' cottage was cozy, the perfect size for a small family. David proudly, meticulously groomed his garden almost every weekend.

In 1933, many people in Palo Alto were certainly more fortunate than the rest of the country. The United States had been struggling to survive a world economic crisis since 1929. The Great Depression had devastated so many families—fifteen million Americans were unemployed at the time, about 25 percent of the country. But most people in Palo Alto seemed to be thriving, or at least maintaining.

Professors and scholars at Stanford University continued to teach classes and conduct research. Endowments suffered, but athletics and academics had expanded. The city relied on the university's faculty and staff to spend money—and they did.

The black smoke billowed from the bonfire. It was a glorious summer morning in Northern California—bright, blue skies with just a hint of warmth. Unlike San Francisco, its Bay Area neighbor to the north, Palo Alto was shielded from the cool summer fog by the Santa Cruz Mountains.

The yard trash slowly cooked. But buried inside the pile was an innocuous piece of metal that refused to melt as it seared beneath the embers. In just a few hours it would become a vital clue, but for now it remained one more piece of junk in David Lamson's bonfire.

Around nine that morning Allene Thorpe Lamson untangled her brown hair with her fingers, gently dividing it into sections and then weaving two long braids. Wrapped in her cotton nightgown, she gazed into the mirror hanging on the vanity in the couple's small master bedroom. Allene was a natural beauty, with a slender figure, pale skin, dark hair, and chocolate-colored eyes, but her most attractive feature was her mind. She had received both a bachelor's and master's degree from Stanford University, an impressive achievement for anyone in the 1930s, particularly a woman. Allene had belonged to myriad campus organizations—a leader in the Delta Delta Delta sorority as well as the women's national journalism fraternity, Theta Sigma Phi. She was president of the Peninsula Women's Stanford Club.

She was a fledgling writer and editor for the university's yearbook, the *1926 Quad,* as well as the *Stanford Daily,* a campus newspaper. As a graduate student she wrote lengthy and deeply researched features, including stories about the school's hefty endowments and the publication of the university's yearbook. Her writing was fluid and engaging—she clearly delighted in journalism.

"In a few short miles one passes from sea level to mountain top, each region abounding in the wild creatures and plants peculiar to it," Allene wrote about Stanford's role as a game refuge.

She was particularly enamored of the gorgeous Northern California countryside. She had moved from her native Missouri several years before, and her surroundings were often featured in her writing.

Inside the yearbook's offices she met David Lamson, the charismatic editor in chief for a popular humor magazine, the *Stanford Chaparral.* They shared so many interests, both brainy students who were engaged in the Stanford community. By graduation Allene had been charmed by the handsome writer, and they were married just a few years later.

Her thirty-one-year-old husband of five years was slim and fit with dark brown eyes and a full head of thick, wavy dark brown hair just beginning to recede at the forehead. Much of the time David Lamson seemed pensive—curious women might have labeled him "intriguing." The outer corners of his eyes drooped just a bit, but his young daughter almost always drew out a sly smile that turned big and bright. He was perpetually charming with friends, which made them a popular couple, much to Allene's delight.

In 1933, David was the sales manager of the Stanford University Press, the school's prestigious publishing house. He had spent a year teaching advertising at the university—a writer with ambition. Allene was an assistant executive secretary with the YWCA, which was more of a job than a calling. The position didn't tap the skills she had earned from her two degrees. It stifled her, but unemployment wouldn't do.

"She needed something to occupy her mind," David explained to a friend. "She was not satisfied to be home."

The Lamsons were a modish couple, both hailing from well-respected families. David was from Cupertino, California—his mother and two sisters lived nearby, one of whom was a well-known physician with her own medical practice. Their friends were some of the most moneyed figures in Palo Alto—there was a chemist with the National Research Council, a metallurgical engineer, a journalism professor, and an attorney. One of their closest confidants was socialite Louise Dunbar, President Hoover's glamorous niece, who cavorted with the city's bluebloods.

Allene gazed in the mirror as she examined the tiny lines on her face, as most women do. She was twenty-eight years old and the mother of a toddler, a little girl with black curly hair she named Allene Genevieve, whom she called Bebe. Allene smoothed her braids, coiled them, and fastened each to either side of her head neatly with hairpins, part of her morning routine. It had been such a taxing night, the last evening of a holiday weekend. She and David had zipped between social events for the last three of four evenings. There was a visit with the Ormsby family on Friday, several bridge games at the Swains' home on Sunday, and dessert with their friends Dr. and Mrs. Ralph Wesley Wright the night before. The Lamsons enjoyed being hosted by friends, intellectuals who challenged their ideas and tickled them with quick wit.

"I would say they were quite happy," remembered Dr. Wright.

But the couple's enthusiastic socializing might have finally taken its toll. After chatting for several hours with the Wrights over dessert the night before, the Lamsons arrived home by eleven with Allene's stomach in knots. Perhaps it was the lemon pie and orange juice that Mrs. Wright served, she wasn't sure. David tried to be considerate; he insisted on lying down in their daughter's nursery at the back of the house so he wouldn't disturb her, which had been their routine for years when she needed rest. Luckily two-year-old Bebe was at a sleepover with David's mother—a blessing, the families would later say.

David reminded Allene that he planned to do yard work the following day; he removed his work clothes, bathrobe, pajamas, and house shoes from the hall closet so he could slip out quietly in the morning. Allene snuggled under the sheets and closed her eyes, but not for very long.

The stomach pain had returned around three that morning when she called his name; there was no need to shout because their house was so tiny. David appeared at their bedroom door in his pajamas. He ran his hand gently across her back to comfort her and then suggested she have a bite to eat.

Soon Allene could hear him collecting things in the kitchen. He handed her a glass of lemon juice mixed with water; then he quickly left and returned with some warmed-up leftover tomato soup and a toasted cheese sandwich. Eating something hot usually lulled her back to sleep, but she had little appetite that night. She nibbled on the crust and took just a few sips of soup.

David returned to the nursery as Allene fell asleep again. The house was quiet now without Bebe; it was almost disconcerting. A silent home meant a respite from the incessant crying of a toddler who had suffered from horrible sinus infections all winter. It had been an exhausting few months for Allene—night after night of coaxing a sick child back to bed with the help of a nursemaid in the little girl's room. David was the one to suggest that Bebe stay with his mother; he also told the nursemaid to take the holiday off so he and his wife could have some privacy. With Bebe sleeping at her mother-in-law's, Allene was in a peaceful home, despite the indigestion.

By nine that morning, David appeared in the bedroom's doorway once again. His shirt was off, his chest was sweaty, and his face was wet after hours of early-morning yard work near the bonfire.

Allene was still feeling poorly, but David had anticipated that. The water from the tub in the next room rumbled through the pipes—a hot bath was waiting for her. David had also prepared a breakfast tray in the kitchen with a bowl filled with Shredded Wheat cereal, a container of cream, and hot water for her morning cup of Postum, a popular coffee substitute made of whole grains and molasses for those who didn't care for caffeine.

David guided Allene down the short hallway to the left of their bedroom. Much of the tiny bathroom was bright white, including the walls, the fixtures, and the tile around the tub. The room was far too cramped for two people, so David gently maneuvered her around the basin; she suffered from notoriously weak ankles.

Allene kicked off her sheep fleece–lined slippers, untied her nightgown, and hung it on the door nearby. David helped her step into the tub, which was now quickly filling with warm water. Weighing about 115 pounds, Allene was a delicate woman even at her healthiest, and her stomach was still bothering her that morning. She hoped that a long soak might move along her recovery—she didn't intend to wash her hair, just relax. She didn't even bother with a bar of soap.

Allene was steady as she lowered herself into the water, while David turned and left the door slightly ajar, stuck on a thick doormat. The tub was about halfway full when she turned the handle and slowly stood up—it was time to begin the day. The doorbell rang, but it might have gone unnoticed.

Suddenly the light that illuminated her bathroom vanished—deep blackness was everywhere. Perhaps she had closed her eyes, just for a bit, but the sensation was startling, as if she was blinded by thick ink. She was breathless, and now there was an aching at the back of her head, stretching from ear to ear. She collapsed.

The outside of the porcelain tub was cold as her body slumped over the side. Her torso dangled halfway out. Her arms hung down. Allene's head tilted toward the tiles of the bathroom floor as one of her beautiful dark braids, which she had so gently fixed earlier, became unpinned and drooped along her left arm to the floor. The ends of her hair were frayed. One of her hands rested on a slipper, which had been lying on the tiles just outside the tub.

There was blood everywhere—even on the ceiling—but she didn't notice. She was limp, dying. Red liquid from the back of her head quickly spilled into the clear water in the bathtub as crimson tentacles reached away from her body. The water slowly turned pink. Dark red streaks slid along the side of the tub. Within minutes, the blood glistened in her hair, soaking the brown strands along with almost every surface of her bathroom.

Allene Lamson's gruesome death would soon attract more attention than her quiet, ordinary life. Her friendships and her marriage would offer morbid fodder for a scandal-hungry press and a politically savvy prosecutor. Most of Allene's friends didn't realize that her gracious smile had hidden some troubling secrets, but soon everyone would know. She was married to a killer—even he had admitted it. And soon newspapers across America would accuse David Lamson of murdering Allene, too. But that narrative would unfurl later. For another few minutes Allene Thorpe Lamson would lie alone, dying in warm bathwater.

For the past three years David Lamson had been a reliably cordial neighbor. His scheduled weekend tasks in the small backyard were part sweat equity, part social hour. Friends peered over their fences and gossiped with one another about colleagues and classes as they trimmed their lush fruit trees—quince, apple, pear, loquat, and fig, among others.

"I hoed," he remembered, "cleaned away the weeds by the blackberry vines, which I wanted to irrigate."

That morning David's task was to trim his artichoke plants in the back garden, not an unusual edict for many husbands who chose to use the holiday as a day to check off their chore lists. He strolled into the garden around seven after having a small breakfast with coffee. The Lamsons would soon be off to the mountains. They planned to spend the summer away from Palo Alto and would be renting out their bungalow for a few months. There was so much to do beforehand. Neighbors watched David navigate the piles of trimmings and weeds. Right before ten, he stopped for a chat with Helen Vincent about simonizing her car.

"I remarked that he was doing more than one thing at a time," recalled Vincent, "getting a sunbath and doing his garden work."

During their conversation a woman appeared in his garden, Julia Place, the Lamsons' real estate agent. She explained that she had two clients with her from San Francisco who might want to rent the Lamsons' home for the summer. David seemed a bit surprised, because they hadn't arranged an appointment. Allene must not have heard the doorbell's ring from the bathtub.

"He said it would be perfectly all right if I would go to the front," said Place. "He would go through the back door and let me in with my clients."

Place and Vincent watched David slip on his shirt and walk into his house through the back porch, while the agent and her clients returned to the front. Less than four minutes passed before an alarming sound came from inside—perhaps a scream.

"I really cannot describe it," Place would later explain to police. "I would say it was hysteria."

"My God, my wife has been murdered!" he cried as he flung open the door.

Julia Place and her clients, standing on his porch, stared at him. He was screaming, his shirt covered with pinkish-red blotches. His hands and face were dripping with water. Much of what happened next became a series of dim memories. He remembered carrying his nighttime clothes down the hall toward the bathroom.

"The first thing I saw was blood on the floor and the next thing was Allene lying over the tub," David said, "her skull fractured."

He cried out and cradled her, smearing blood across his shirt. She wasn't responding to his voice. He laid her down again and dashed down the hallway, leaving his footprints in her blood along the way. Allene sank back into the tub and hung over the side.

"Of the rest of that morning I remember mercifully little," David said. "It is as if the shutter of my mind opened now and again to photograph a scene, leaving a series of isolated impressions with blank gaps in between."

He begged the real estate agent to come inside.

"Get the police to find the murderer!" David screamed.

He ran to the bathroom again, crying wildly and staring at his wife as he held her again. One neighbor said she could hear his screams from one hundred yards away.

"Some of the things I remember most vividly are matters of no possible importance," David remembered. "A friend's voice urging me to come away, to let my wife go from my arms."

He would later learn that his neighbor Mrs. Brown found him kneeling by Allene's body, crying. She led David toward the nursery before he fainted and collapsed. A horrible scene would later haunt him.

"The glimpse of a neighbor's face, twisted with pity and horror, emerging distinct from the blur of faces that filled the house," David said.

He ordered Mrs. Brown to call his sister the physician . . . and the police. It was 10:10 a.m. when Palo Alto's chief of police and several officers rushed into the cottage. There were now more than a dozen people inside the tiny home. The chief spotted Mrs. Brown holding a bloody towel and scolded her for inadvertently ruining forensic evidence.

"She was down, cleaning up something off the floor," said Chief Howard Zink. "I told her to stop wiping up the blood, that everything must be left as it was, for evidence."

Eight officers had responded, and soon each was interrogating David. A photographer snapped pictures while Allene Lamson lay on display—her naked body was partially draped over the tub for almost two hours. Strangers stared and whispered. The coroner noted several lacerations and contusions to the back of her head. One investigator shoved his hand in the tub just inches from her body and declared that the water was still warm. The doctors tested for rigor mortis, the stiffening of the limbs and joints that happens about two hours after death. They could still rotate her head, and the autopsy later concluded that Allene had died about an hour earlier, sometime after she climbed into the tub.

"Who could have done it?" David cried. "No one had anything against her."

The house was a nightmare for investigators. Allene's blood had been transferred to almost every corner of her small home. The pathologist, the undertaker, officers, and countless neighbors had all shuffled through the scene, along with David Lamson and the real estate agent. There were large pools of blood in the bathroom, splashes in the hallway, red footprints leading to both bedrooms, sprays containing hundreds of droplets on each bathroom wall, and smears wiped on doorknobs. Reconstructing the scene would be arduous, even for more experienced detectives.

Officers peered down at the bathroom floor. It seemed improbable that a petite woman could be responsible for so much blood. Doctors guessed that about half of her blood had drained from her body, about two-thirds of a gallon. Some of it was diluted by water from the bath. Some of it was arterial—blood that had sprayed directly from the body and had not mixed with any other fluid.

After two hours Allene was hoisted onto a stretcher—she spilled more blood along the route to the front porch. Her neighbors were awestruck. It was a ghastly, abrupt ending for an accomplished woman who had commanded respect from the time she strolled onto Stanford's campus. Allene's death would soon become even more troubling, particularly for her husband. The police chief eyed David as he answered questions.

"Ten minutes after the deputies arrived they were accusing me of murdering my wife," David said. "Two hours later I was in the jail in San Jose."

Less than twelve hours after Allene's death the press happily latched on to the story. Subscribers to the local newspaper, the *Santa Cruz News,* found a piece with the titillating (and long) headline "Prominent Young Palo Alto Woman Is Found Dead in Bath Tub with Gaping Hole in

Back of Her Head." Hundreds of newspapers across America had picked up the report by the end of the day.

"Sheriff William Emig expressed belief she had been slain," the copy read. "David Lamson . . . could offer no motive for his wife's death."

David seemed to agree with the sheriff that someone had broken into his cottage, perhaps a robber, and killed his wife while she bathed—there was no other explanation. The story was read by a remarkably large number of American readers who begged to be teased by the media. A year earlier, the famous aviator Charles Lindbergh's twenty-month-old son was kidnapped for ransom from the family's mansion in Hopewell, New Jersey. For two months federal agents led a massive manhunt for the baby before his body was found in the nearby woods.

Federal agents tracked thousands of leads in the Lindbergh mystery. Each new detail prompted a media frenzy and sold millions of newspapers, but by the spring of 1933, there were few updates, so readers were eager for another scandal. Now newspaper editors offered up the mystery of a prominent university academic turned wife killer as the next big headline. With each twist readers demanded more details, preferably lewd bits of gossip disguised as facts.

"Mystery Man Adds New Theory Puzzle," declared a headline printed two days after Allene's death. A university student had spotted a "shabbily dressed stranger loiterer near the vine-covered campus cottage." The witness said the man was lurking by the home early Tuesday morning when Allene died. Not credible, according to police, because David Lamson was the murderer. The couple's friends weren't convinced; the disquieting rumor snaking its way around the upscale neighborhoods on campus was that there was a killer stalking the upper-class houses along Stanford University's Faculty Row.

"Guest." That's how Santa Clara County sheriff William Emig described David Lamson's status in the jail in San Jose, California. He was not arrested or charged, but he *was* being held while investigators scrambled to sort through evidence. The police knew they were running out of time, because Lamson's attorney was complaining loudly to the media about false arrest allegations.

In his three-piece brown tweed suit, David Lamson sat at an old wooden desk with a pen and paper inside his jail cell, less than twenty miles from his home. Scribbling near an oil lamp, he seemed more like an academic than a prisoner. Photos of Allene and Bebe were taped to the walls. David's cell was on the third floor—the only room on that floor. He looked at his gold wedding ring. He had just fifteen square feet in which to pace, to fret about his trial. He often looked at the iron door with its large bolt. The walls were stone and steel, making the echoes inside the tall building almost unbearable. Lying alone on his cot, David thought about his daughter, Bebe, who was now living with his sister Margaret. His two sisters and their mother all steadfastly believed in his innocence.

At his desk, David Lamson considered his moderately circumscribed life. He had graduated from Stanford University in 1925 and was immediately hired by the academic press—a promising job that would become a successful career. He had married Allene Thorpe three years later, and they bought the bungalow on campus the following year; their daughter, Bebe, arrived the year after that. David had crafted for himself a predicable but pleasant life of socializing, family activities, and rewarding work. But as in most marriages, Allene and David had struggled with problems. And he kept secrets.

A San Jose jail cell would be his home for now as he awaited trial. He peered toward the street from his window and gripped the vertical bars. He could glance at Allene's picture, wistful for days when she was alive. He had few emotions left, and grieving might not have been one of them, because he was scared for his own life. His defense team, some of the most talented attorneys in the state, watched him. They would hire only the best experts, they assured him, men who could expose weaknesses in the evidence gathered by the state's investigators; those experts would set him free.

"It never occurred to any of us that anything but an acquittal might result," David said.

Back at the Lamson house, a brunette was prone, stretched across the tiny bathroom with her face pressed against the floor. The tops of her knees rested on the edge of the white tub. Her head rested just beneath the basin. Her arms, bracing against the floor, held up her body.

It was June 20—weeks after Allene Lamson had been found dead—when Edward Oscar Heinrich (Oscar, to his friends) waited in the doorway, peering at his "model," his assistant's wife, who had reluctantly agreed to play the distressing role of "corpse" for photos. She slowly stood up, readying herself for a new position while Oscar adjusted his small, round wire-rimmed glasses. It was his second trip to the Lamson house in a week. He jotted down notes and glanced at the reddish-brown spots on the wall.

"The door is liberally spattered below the glass," he scribbled in his journal. "On the door jamb the drops show a projection southerly and upward which carry back to this same point."

Hovering near Oscar was a fellow in a smart, dark three-piece suit sans jacket—Palo Alto criminalist George A. Weber. Oscar could feel Weber's gaze. Oscar snapped another picture of Jean Weber with her arms flung over the side of the bathtub and her head tilted downward, a replica of Allene Lamson's pose in death.

Oscar Heinrich had read books by European investigators about how a body might release its blood when impacted. And he had honed his technique on earlier cases, introducing perhaps the first blood-pattern analysis (BPA) testimony in America during a California murder trial in 1925. He was a professionally trained expert in a multitude of forensic sciences, including chemistry and biology, unlike many of the charlatans he disputed in court. By 1933, he was more experienced in bloodstain-pattern analysis than virtually anyone in the United States.

David Lamson's case was stymied by a mismanaged crime scene. Palo Alto police couldn't untangle the rumors—there were unfounded whispers of an affair with a writer, a tryst with his daughter's nursemaid, and violent arguments over sex that resulted in Allene ejecting David from their bedroom most nights. The local cops didn't have the skills to focus on the facts, so they chased phantoms for two weeks. Much of the forensic evidence was lost in a chaotic scene on the day of her death.

A forensic expert's first assignment is to preserve the evidence by securing the crime scene. David Lamson's home had been contaminated by loads of people poking around the house. Oscar considered the case—his good friend August Vollmer, the former Berkeley police chief and now a college professor, would be helpful here, along with the librarian John Boynton Kaiser. Oscar had called on Vollmer's advice for many cases because he was a cop with incredible investigative instincts who believed that educated police officers could outthink even the sneakiest criminals. Oscar was always pleased when they were able to work a case together.

"A precise little man," observed one reporter—Oscar enjoyed an international reputation as a criminalist who could reconstruct a crime by collecting hidden forensic clues, processing them in his lab, and testifying as an expert witness. Most times Oscar was a prosecutor's savior during tough trials. He was the scientist who had erected the nation's first private science laboratory in 1910—America's earliest general forensics lab. By 1919 there was another facility, one specifically for New York City's toxicology cases, but Oscar's laboratory was equipped to test all aspects of forensics—the lair of a real-life Sherlock Holmes.

Oscar Heinrich's most high-profile cases had been splashed across American newspapers for two decades. Suspected killers confessed when they spotted his name on their case files. Journalists around the world had gleefully compared him to the most famous investigator in history—the greatest detective who never lived.

"So here is the inner sanctum of Sherlock Holmes," a newspaper reporter once quipped as he wandered around Oscar's Berkeley lab in Northern California.

"Not Sherlock Holmes," Oscar snapped, shaking his head. "Holmes acted on hunches. And hunches play no part in my crime laboratory."

After hours of work, Oscar finally summoned the journalists milling outside the Lamson home. He rarely disclosed many details to reporters because he didn't trust most of them, but the public demanded an update.

"I have discovered enough evidence to warrant my staying on the case," Oscar explained. "All action in this case took place in the bathroom, and I can reconstruct it in detail. I was delighted to discover a number of important clues overlooked until now by both the prosecutor and defense."

At the end of the day Oscar had gathered all he needed. His hypothesis was sound and his suspect was near, he believed, and he could prove it using pieces of string and hundreds of dried blood drops that freckled the bathroom wall. The height and angle of the blood, along with its trajectory, would reveal the angle of impact. Allene's every movement, Oscar theorized, would result in a very specific pattern. The strings, the protractor, the calculations, and the drips of Allene's blood would solve her death.

"*X* marks the spot," he said as he turned to assistant George Weber.

By 1933, Oscar Heinrich had unraveled countless violent crimes that seemed too puzzling to crack. His notoriety was impressive, but his record was not unblemished. His work over the past decade and his growing reputation for solving so-called impossible cases gave many police officers and jurors confidence in his formidable abilities. But he had also made serious miscalculations in cases along the way, and his occasional aloofness on the stand sometimes stood in the way of getting the convictions he had so fervently pursued. The idea of solving crimes with forensic science was still such a new concept, and he was constantly fighting the perception that his techniques were "unproven," "untrustworthy," or "unreliable."

Oscar was convinced that Allene's death could be solved with the cutting-edge forensic tools that he was pioneering in his lab. A person's life was in his hands, a weight he had carried many times before. Looking back on his own volatile career—his missteps at trials and his vicious fights with other experts—it pained Oscar to realize that there might, in fact, be room for doubt.

But before we can understand the conclusion of Allene's case, we must first go back to the beginning—not of her story but of Oscar's. And for Oscar, the beginning was a series of formative, and tragic, events in his childhood that would influence and inform all the work that would follow over the next seventy-plus years of his storied career.

2.

Genius: The Case of Oscar Heinrich's Demons

"They say that genius is an infinite capacity for taking pains," he remarked with a smile. "It's a very bad definition, but it does apply to detective work."

—Arthur Conan Doyle,
A Study in Scarlet, 1887

Edward Oscar Heinrich watched his mother mill around their small kitchen in Tacoma, Washington, as she collected breakfast dishes. It was October 7, 1897. The cups clinked in the sink. The sixteen-year-old slowly ate his meal. Reflected in the drinking glasses was a slight woman with an attractive face, wide-set eyes, and dark hair—the strongest, most steadfast person he would ever know. She was his moral guide, and by the end of that morning, he would become her savior for the remainder of her life.

Oscar had watched his mother suffer through much of his childhood. Albertine had been just twenty years old when she and twenty-eight-year-old August Heinrich, both natives of Germany, married in the Trinity Lutheran Church in Wisconsin. A year after having a little girl, Adalina Clara, Albertine gave birth to a boy they named Gustav Theodor Heinrich. The baby lived only a month, a tragedy for the young family. Another girl, Anna Matilde, came shortly after Gustav's death, and soon Oscar arrived on April 20, 1881.

He would be the Heinrichs' last child; and to honor the brother he never met, Oscar later named his eldest son, Theodore, after him. Oscar later remembered his mother warmly—she was a solid, secure presence in the young man's life, and he admired her moxie and her deep sense of duty to their family. She leaned hard on both attributes, because from the time she was married, the family struggled with money.

"We kids earned our pennies by gathering old whisky bottles outside the factories and selling them," Oscar remembered. "We were paid a penny for small ones, two cents for larger ones. That was our only source of spending-money." There was never enough money to go around, but even for a young immigrant family there were always ways to scrounge up a little cash. And Oscar was nothing if not resourceful.

When Oscar was nine, August moved the whole family out west to Tacoma, Washington, for better opportunities promised by the newly finished railroad. But life there wasn't much easier, and Oscar grew frustrated as he found himself surrounded by privileged and entitled children in the then-booming lumber town. His father wasn't able to provide him with an allowance, so Oscar became determined to earn his own money. He was soon hired for a newspaper route, a lucrative job, but one that took him to the city's red-light district.

Oscar had never strayed much beyond his family circle, and his forays into Tacoma's underbelly to sell news to the area's less reputable denizens were enlightening. He kept his eyes open—and his sense of decorum intact.

"Our family's solidarity and my mother's teachings served me well," he said. "When I approached women in saloons and offered them my papers, I always had my cap in hand. They seemed to respect me."

That entrepreneurial flair, linked with his own love of the written word, led young Oscar to embrace journalism of all kinds. Not content to merely hawk the news, the teenager reported and penned a newspaper story about the new game of handball to the *Tacoma Morning Union* in 1895, when he was in eighth grade, moving from delivering the news to actually writing it. But his moneymaking endeavors weren't just an adventure: they were increasingly a necessity. That same year, patriarch August Heinrich lost the family's savings during a recession, and fourteen-year-old Oscar was forced to leave high school for a few months to take a janitorial job in a pharmacy, an entry-level post that would eventually serve as the foundation of his career.

Oscar read constantly in his off-hours: English literature, scientific tomes, and language primers. He also began tinkering with fiction writing—rudimentary, silly detective stories. He liked making money in his janitorial job, but he knew he wanted more of an intellectual challenge out of his work. Luckily, his family's finances improved somewhat, and he returned to school less than a year after withdrawing with dreams of moving overseas. But when Oscar confided his plan to his father, August Heinrich stared back and issued a stern warning that portended difficult times ahead.

"You have no brothers," he cautioned. "If anything happens to me, it's up to you to support your mother and sisters."

With broad shoulders and callused hands, Oscar's father worked as a skilled carpenter in his woodshed behind their Tacoma home, pushing saws and driving nails into boards for hours. But the forty-nine-year-old struggled to find steady work. August was handsome, even with scruffy brown hair and a ragged beard that framed his face—the antithesis of the cultivated public image Oscar would later strive to achieve (and would always demand of his own sons). He and his mother thought that things were improving with his father and their fortunes. But the reprieve was short-lived.

Just after six on the morning of October 6, 1897, August pushed his chair back from the breakfast table and picked up his kit of tools, informing his son and wife that he would be leaving for a carpeting job on C Street. He glanced toward his woodworking shed at the back of the house and wished them a good day. As he walked through the back door, Albertine had no inkling that she would never see her husband alive again, no idea that at forty-two years old she would be abandoned with a teenaged son and two unmarried daughters.

After August disappeared in the backyard, Albertine noticed that her husband had left behind his dinner bucket. She walked swiftly toward the shed and swung open the building's door. She screamed so loudly that most of their neighbors heard.

In less than a decade, Oscar would become one of the greatest forensic scientists in history. But at that moment, standing in his parents' kitchen, he was just minutes away from seeing the first of many, many corpses in his career. The scene would plague Oscar until his own death, a ghastly reminder of what might happen if he surrendered to his own flaws.

The teenager sprang up from the table when he heard his mother's wailing; he raced through the backyard and stood in the doorway of his father's woodshed. Oscar looked upward as his mother collapsed. His father was hanging from a wooden beam near the ceiling with a window cord tied around his neck, dead from suicide.

As his mother sobbed, Oscar did something extraordinary for a sixteen-year-old boy. He gently led his mother to the kitchen and settled her in a chair so she wouldn't faint. He phoned the police, retrieved a knife from the kitchen, and returned to the woodshed; Oscar climbed the stepstool that his father had used to slip on the noose. August's body shook and swayed from the violent hacking of the knife. It finally dropped to the ground. Exhausted, Oscar dragged him to the house. Soon the police arrived, followed by local newspaper reporters, who collected details on the family tragedy.

"Suicide at Glendale," read the *Tacoma Daily News*. "August Heinrich Hangs Himself in His Workshop."

"No reason for the deed can be discovered," read the copy, "and his wife and family, as well as many of his acquaintances, are at a loss to explain the cause of the suicide."

But Oscar's family knew the truth: his father had been distressed over finances for years, yes, but he had also been plagued with an ongoing darkness that clouded his life, one he had always struggled to overcome. The coroner determined that August had died from strangulation, not a broken neck—a more prolonged, painful death. It was an agonizing end for the flawed father Oscar loved so much.

Over the next six decades Oscar secretly fretted that he might also suffer from his father's same anguish, his same craven weaknesses. But that fear also shaped his future—it spurred on his determination and helped craft his acumen for controlling every aspect of his life. Oscar transformed frustration into resilience. His deficiencies, such as his obsessive compulsions, became attributes . . . until they threatened his career, his family—even his life.

"Among my earliest recollections, the most prominent are those of the brutal ways in which I have been robbed of all my illusions," Oscar wrote his best friend, John Boynton Kaiser. "It stirs resentment, suggests revenge and breeds caution."

After his father committed suicide in 1897, sixteen-year-old Oscar was immediately assigned immeasurable responsibility, almost more than he could bear. Reporters knocked on his door, demanding answers about his father's death, the first of many secrets Oscar Heinrich would protect until he himself died. His father's grim fate would haunt Oscar for decades, testing his relationships with his children and challenging his own mental health.

When he became the patriarch of the Heinrich family, there was no hope of returning to high school for his final two years. He studied at night to become a pharmacist, a steady career for a young man who needed to support his family. By the turn of the century, pharmacists were still sometimes called apothecaries. They dispensed medications, prescribed remedies, and even gave some treatments that were difficult to self-administer, like enemas. Going to a pharmacy school wasn't needed to take the state board exams, but apprenticing under a licensed pharmacist was required, usually for at least a year.

The pharmacy curriculum of the early 1900s leaned heavily on chemistry. It trained a pharmacist not only to prepare medications but also to practice clinical chemistry, which was the analysis of bodily fluids, like conducting a urinalysis. The training was Oscar's first step toward becoming a forensic scientist—a career he never initially intended to follow.

The teenager couldn't afford formal pharmacy classes, so he depended on his innate ability to understand the context of the texts and the math behind dispensing the correct amounts of medication. He relied on his memory and his compulsion to stockpile useful information. When the eighteen-year-old passed his pharmacy state exams, he seemed to be the only one who wasn't surprised as he slipped on a white coat. Inside the Stewart and Holmes Drug Company he studied drugs, poisons, chemicals . . . and human nature.

"A drugstore is a veritable laboratory in behavioristic psychology," he said.

He watched male customers slyly leer at women in the store. Some tried to bluff Oscar to secure more medicine without a prescription. Others were desperate, clearly addicted to the medicinal alcohol he could access for them. And a few customers turned mean, even threatening, if they didn't get their way—when they thought no one was listening.

"I learned what people do in secret," he said with a smile.

Pharmacy work also offered him another reward—eight years of excellent training in the valuable skill of handwriting analysis.

"I had doctors' prescriptions to decipher," Oscar explained. "And doctors are the worst writers in the world. Right then I started in to qualify as a handwriting expert."

For almost a decade, he watched the other pharmacists quickly calculate the formulas for medicine—and he envied them.

"I was impressed with the difference between the caliber of my work and that of college-trained men," he said. "Those men were far away from me as technicians."

He desperately wanted to be a skilled chemist. He needed to secure a well-paying, stable job in Tacoma to help his family. He hoped to spend his life bending over beakers in a lab, but to do that, he had to go to college, and it wouldn't be easy. Even though he had his pharmacy license, Oscar lacked a high school degree—and despite his years of work, he'd managed to save very little money. Still, he was determined, and he'd heard about a special program for nontraditional students like himself at the University of California at Berkeley that sounded like just the ticket for a driven (but uncredentialed) student like Oscar. With just $15 in his wallet, the twenty-three-year-old planned to embark upon the next phase of his education.

But just three hours before he was scheduled to board the train to Berkeley, a disaster: he received a letter from the university that said he had missed the entrance exams to become a special student by two weeks. The details of the mix-up are lost to history, but true to his enterprising nature, Oscar was determined to achieve his goals. He hopped the train anyway and turned up at the registrar's office in Berkeley, demanding to be admitted.

"When I presented myself the Recorder of the Faculties listened to my story, looked me over, then told me to go up to the Chemistry College and go to work," Oscar would later tell his son.

It didn't happen right away, but finally Oscar's perseverance and intelligence convinced admissions officers to allow him to join the freshman class of the College of Chemistry as a special student in chemical engineering. It was a good gamble. He quickly became invaluable to his professors as a laboratory assistant in quantitative analysis, and then as an assistant instructor in physics and mechanics. He studied medicine and attended law courses, then took classes in sanitary engineering—a discipline that used science and math to improve sanitation along with the supply of safe potable water.

His Bachelor of Science degree taught him how to uncover nearly invisible clues and become a specialist in chemical jurisprudence who could detect poisons and identify mysterious stains. Amid furious bouts of studying and teaching, he managed to make time to woo a pretty co-ed, a calming companion during his transition from student to independent man.

Marion Allen and Oscar Heinrich met on campus at the University of California at Berkeley as students. They were both involved in Greek life; Marion was prominent with Delta Delta Delta, while Oscar had been elected to the Mim Kaph Mim chemistry honor society and the Acacia club, a social fraternity founded by undergraduate Freemasons. They took chemistry classes together, though Marion never seemed to use her degree for a profession—being married to E. O. Heinrich was likely challenging enough.

Their friends playfully nicknamed Oscar "Heinie," just to gig the fastidious student, who could be a know-it-all. He seemed stoic much of the time, aloof even in his youth, but those who were close to Oscar, like his wife, knew that he was also witty and loving.

"It may seem delightful," he wrote to Marion about the fancy Mayflower Hotel in Washington, D.C., "but without you it is all sheer near-beer."

Oscar and Marion were married at her parents' home in San Francisco shortly after he graduated in 1908. They immediately moved to Tacoma, where his mother lived, and Marion gave birth to Theodore two years later. She was a homemaker, bright and social—a good writer who kept up with national news, but she didn't seem to talk much with her husband about his job. Much of the letters involved neighborhood gossip, the boys (a second son came a few years later), and household finances.

Opportunities came in quick succession after Oscar's wedding to Marion. In Tacoma, he took a job as a chemical and sanitation engineer for the city, where he dealt with paving, bridges, and the development of water and power plant construction. He inspected reservoirs, tunnels, dams, and bridges. He studied the city's sewer and irrigation systems and then designed two chemical plants. And soon Oscar took the job of city chemist, a position that required him to do quite a lot of investigative work with the coroner and the police in cases involving complex chemicals.

But he was frustrated both with the pay and the lack of equipment, so in 1910 he resigned and opened his own private industrial chemical lab called Heinrich Technical Laboratories, a company that helped develop and manufacture products and create processes for clients. But Oscar's work with the police and the coroner continued to stoke his interest in forensics, so he received more criminal cases from the city and the public.

Oscar quickly realized that chemistry faced limitations in criminalistics—there were so many clues to miss without the right training. He expanded on his expertise, spending nights poring over books. He began studying poisons, fingerprinting, geology, and botany—the studies of a forensic generalist. Inside his lab in Tacoma he solved bizarre cases, like the mystery of a poisoned lemon pie sent to a man who became deathly sick after a small taste. Oscar examined the finely sugared crust under his microscope, which proved to contain poison crystals. But there was one big case that established him as a forensics expert in Tacoma shortly after he opened his lab.

He examined the body of a woman found slumped behind a stove in her kitchen. A revolver lay next to her. Investigators labeled it a suicide, but Oscar wasn't so sure. Police led him to a wall behind the victim—there was a hole with a bullet hidden deep inside. But Oscar examined it with his tweezers and found dust inside. A bullet had not passed through it recently. He squatted on the floor and examined every inch of the wall and found two important clues: a small indentation with a bit of lead inside and some washed-out blood on the wall nearby. Now his training in ballistics became practical.

He used string to trace the trajectory of the bullet from the wall to a spot where the shooter was standing—it was far away from the woman's body. He used another string to trace the path where the woman was shot near the wall to where she staggered and fell behind the stove. It was murder. Police arrested her husband, and Oscar was the main witness, securing a conviction. The young, serious-faced chemist was now a star forensics expert.

Soon his constant need to be challenged required a change of career, and it came at a fortuitous time. Through a mutual friend, Oscar would meet an investigator who would help him dissect some of his toughest cases, a cop who was also beguiled by science—American Sherlock's own Inspector Lestrade.

By the late 1910s, August Vollmer was something of a luminary in the Bay Area, Berkeley's first police chief and the man who had reformed police methods nationwide. He would be later nicknamed the "father of modern policing," a revered figure in law enforcement.

In cities at the turn of the century, criminal investigations were predominately "solved" on hunches—the instincts of experienced but ill-equipped detectives who sniffed out suspects based on motive, a dangerous guessing game that leaned on mistaken intuition. A gun and a badge were the only requirements to be a police officer, and a hard spray from a rubber hose was still a common way to make suspects talk. American investigators hoped for clues, hunted for witnesses, and bullied suspects into false confessions based on little or no evidence. It was madness for anyone caught in the legal system, particularly for minorities and immigrants.

August Vollmer demanded reform, and his changes were swift and far-reaching. He created one of the nation's first centralized police records systems, and he was the first police chief to require that cops receive college degrees. Vollmer outlawed the third-degree approach to abusing suspects to get confessions. He argued that a type of truth serum called scopolamine, developed in the early 1920s, was more effective. Vollmer called it the fourth degree. Scientists now know that there is no current drug that can effectively enhance truth-telling, but Vollmer's endorsement of the serum was enough to encourage police departments across the country to reduce their own brutal interrogation methods.

Vollmer trained and then hired African American cops and female officers. He was the first chief to create motorized patrol, buying motorcycles and cars for his officers so they could cover wider areas. He believed in compassion, but he was often criticized for being too lenient on petty criminals. Vollmer gave sound advice to all his rookie cops.

"Your main job as a cop on the beat is not to make a lot of arrests, but to help prevent crime," Vollmer told them. "The best way to do this is to start with the children. Make friends with them. Guide them towards law abiding citizenship. Show them that the law is their friend not their enemy."

But it was Vollmer's emphasis on science that really endeared him to Oscar Heinrich, whom he met through a mutual friend. They had both read the same books by European forensic experts—they spoke the same scientific language. In his own department Vollmer insisted on using forensic evidence like blood, fiber, and soil to solve crimes, and with Oscar's guidance, he created the nation's first police lab, one that tapped the criminalist's expertise in all forms of forensics. Their philosophies on education were symbiotic.

In 1916, Vollmer recruited Oscar to design an innovative program for police officers, America's first "cop college." They wrote back and forth about courses, adding some and scrapping others. They discussed faculty hires, classroom locations, and the most updated methods in forensics. They exchanged syllabi and dissected each line, hoping to improve the other's descriptions for clarity.

"Your suggestion for instruction in library work is timely," Vollmer told Oscar, "and will be taken advantage of."

As Vollmer and Oscar crafted a degree structure together, their friendship grew.

"The course extends over a period of three years, the first-year courses being—physics, chemistry, physiology, anatomy and toxicology," explained Vollmer. "The second year surely requires a college education—criminal psychology, psychiatry, criminology, police organization, methods and procedure. The third year completes the course with microbiology and parasitology, elementary and criminal law."

The next year they launched the School for Police at the University of California at Berkeley, and Oscar began teaching the nation's first criminology classes. Vollmer was incredibly grateful.

"Your outline of lectures is very comprehensive, and should prove of great value to all students of criminal investigation," he wrote Oscar.

Vollmer and Oscar subscribed to the same philosophy: educated police officers were America's most valuable asset in law enforcement. Vollmer hired experts like Oscar to teach elementary law, criminology, and forensic sciences like fingerprinting, handwriting analysis, and ballistics. Police departments across the country sent their officers to UC Berkeley. They listened to Vollmer's lectures on police organization, administration, methods, and procedure. Students lauded Oscar's courses on chemical jurisprudence, judicial photography, applied optics, handwriting analysis, and the use of chemistry and physics in evidence.

He was a demanding, dynamic educator who boasted of his packed classes—though he refused to be called "professor." He believed that while he *had* earned the title, it was a bit too stodgy for a worldly criminalist who tracked killers. Oscar's unique gift to his students was his breadth of field and lab experience coupled with an astounding amount of book knowledge.

"I'll take you all from a thrill to a shudder," Oscar promised a crop of fledgling police officers. "But a chess board in police station is more valuable than crime books. There is nothing that I know of in English that will quickly enable a man to see situations and analogies."

Only an educated, scientific investigator could catch criminals, Oscar and Vollmer believed—everyone else had to pray for luck.

"The investigation of crime is merely a special case of the study of behavior," Oscar told his friend. "Success in it depends upon the development to a high degree of the powers of perception and to an almost equal degree the powers of memory and ratiocination [reasoning]."

His students, who were exclusively police officers at the beginning of the program, were frequently skeptical. Luckily, August Vollmer and Oscar Heinrich were intellectual equals, both committed to crime solving and teaching future detectives. They forged a lifelong partnership, a union of trust and support. They became close friends, bonding over similar upbringings. Like Oscar, Vollmer was born to German parents, and he had also lost his father at a young age.

They would work together on many cases, including a murder trial in San Francisco involving a gang that planted a suitcase bomb during a parade, killing nine people. Oscar helped establish that the bomb parts found at the scene matched the material found in one suspect's room. Vollmer helped Oscar match handprints in the trial of a man accused of murdering his wife by shooting her in the back of the head. They leaned on each other countless times over the years. But soon their loyalty to each other would be tested over a coveted position teaching alongside legends in forensics.

By 1917, teaching young law enforcement officers wasn't enough to satisfy Oscar's deepening interests in policing. At the age of thirty-five, he was named the police chief of Alameda County in San Francisco Bay despite having no formal experience in law enforcement. In California, he restructured the police department, and he dabbled in forensics by studying handwriting with Thomas Kyka, the famous trial expert.

Each of Oscar Heinrich's careers required perfection, orderliness, and a rigid attention to detail. They demanded an incredible ability to synthesize thousands of crucial pieces of data and then recall where they were all stored and how to use them. He hoarded information in every corner of his laboratory—along bookshelves, atop desks, and inside cabinets, all classified to his standards. Oscar's measured methodology was a remarkable system that provided his students and colleagues with a blueprint for how to organize evidence to solve a crime. But the more stress he endured, the more compulsively he organized—his lists and charts became a salve for the anxiety and insecurity he had felt since childhood.

From the first day he had slipped into the white jacket of a pharmacist two decades earlier, Oscar had evolved into an efficient, potent, and hyper-organized investigator, a criminalist who could survey a crime scene and determine which of his myriad tests might solve the case.

By 1918, he and his wife, Marion, had two sons, eight-year-old Theodore and four-year-old Mortimer, and he needed a new challenge. In January, Oscar defeated more than one hundred other applicants to become Boulder, Colorado's first city manager and commissioner of public safety for about $5,000 a year, but the job lasted only a year. When his mentor, Thomas Kyka, suddenly died in San Francisco, Oscar returned to take over his questioned-documents business while teaching at UC Berkeley with Vollmer. And he strengthened an already close relationship with a brilliant cohort, a reference librarian from his days in Tacoma—his own Dr. John Watson.

Oscar vented often about other competitors, but not to his wife or even to his colleagues at UC Berkeley. He griped in hundreds of letters to John Boynton Kaiser, his closest confidant.

"His propaganda is partially for the purpose of personal publicity rather than the scientific value his alleged discoveries may have," Oscar complained of one rival expert.

Oscar and Kaiser had written almost weekly since they met in Washington State in 1914. Kaiser wasn't just a close friend—he was also an outstanding researcher and a published author. The thirty-four-year-old had written well-respected guides for librarians on organizing municipal and legal documents, on the national bibliographies of South American republics. And he enjoyed criminalistics, writing an article for the *Journal of the American Institute of Criminal Law and Criminology* about people who bought and sold children.

Kaiser received a master's degree in library sciences from the illustrious New York State Library School in Albany in 1917. After school, Kaiser was appointed librarian at the Texas State Library before moving on to a bigger position at the University of Illinois and then finally to Tacoma. The researcher became Oscar's meticulous adviser, a font of material who sent him hundreds of books paired with loads of advice over four decades. Kaiser often happily played the role of problem-solver for Oscar.

The librarian was a smart dresser, slightly portly with dark thinning hair and a toothbrush mustache, a fussy but brilliant writer and bookish amateur detective. He enjoyed flagging useful articles for Oscar, like a piece on how to identify an authentic document.

"On page 363 of the LIBRARY JOURNAL is a list of ten tests by which the original may be distinguished from any of the reprints," wrote Kaiser. "This may be of some value to you at some time."

Oscar utilized Kaiser as a sounding board for personal problems, health queries, financial concerns, and criminal cases. When he lectured to students or appeared on the witness stand, Oscar remained unflappable and confident, the embodiment of a savant in criminalistics. But Oscar's public bravado masked private self-doubts and insecurities that he rarely revealed to anyone but Kaiser.

"Sometimes I enjoy your insistence in thinking I am a great detective," wrote Oscar, "and other times I wonder where you get the idea. I remember that I told you in one of my recent letters that I never talk to people. I simply look them over."

Kaiser advised Oscar on books and references throughout some of his most challenging trials. The forensic scientist had first appeared in international newspapers about five years earlier, when he cracked enemy codes for the American government during World War I. In 1916, Indian nationalist groups attempted a rebellion in India against Britain, a plan dubbed the Hindu Ghadar Conspiracy. Oscar spent months with tutors, learning the three distinct Hindu dialects that were represented in the letters. It wouldn't be enough to just translate the codes; he wanted to learn the nuances of the language to decipher the messages and prove their authorship using patterns in word choice and writing style—profiling.

With the help of more books from Kaiser, Oscar analyzed the handwriting and typewriting on hundreds of documents, and he used chemicals to test ink. Working alongside investigators with Scotland Yard and the American government, they finally solved the case, a conspiracy that involved more than thirty people. By the end of the war Oscar earned the rank of captain in the U.S. Engineers' Reserve Corps and the admiration of federal investigators.

In 1920, he testified as an expert in handwriting analysis in heavyweight champion boxer Jack Dempsey's draft-dodging trial, though Oscar never took the stand, and the "Manassa Mauler" was acquitted.

Oscar used hair comparison analysis to secure the conviction of two soldiers accused of beating to death a taxicab driver in Salinas with a blackjack (a type of thin club). And with virtually each case, Oscar offered Kaiser a tantalizing, secret glimpse behind the process of forensic investigation.

"I am enclosing a print showing the hairs as they were found on the blackjack," Oscar wrote to Kaiser. "I was able beyond doubt to establish these hairs as having come from the head of the taxi man."

The pair exchanged books, photos, and opinions. Both were insatiable readers and news consumers. And they shared another interest: studying and collecting stamps. When Oscar was a kid, he had enjoyed deciphering the dates on used stamps as a hobby.

"Whenever you are ready send the 3Cent Lincoln which you have mentioned," Oscar wrote the librarian.

Kaiser and Oscar often sparred intellectually, but they deeply valued their friendship, revealing details about their lives that they didn't dare tell others, even their wives—*especially* their wives. The librarian recognized that Oscar's life as "star witness" was complicated because his friend had been agitated over finances for as long as they had known each other. In 1921, Oscar was expanding his forensics lab, a large space tucked on the ground floor of his three-level house in the hills of Berkeley. With large glass windows overlooking San Francisco Bay, Oscar was finally building a proper laboratory, his first office since he had left Tacoma four years earlier.

"I now use three rooms," he told Kaiser, "a combined laboratory and camera for photographic enlarging and photomicrographic work; a separate fully equipped dark room; and a room for my secretary and auxiliary library."

But the expansion was costing the Heinrich family loads of money, and with two little boys and a wife to support, Oscar worried. The 1920 recession had left his business light, so he agonized over bills, much like his father had decades earlier.

"When I bought the car recently it put such a crimp in the cash account that I will have to work all winter to recover," he told Kaiser.

Oscar was desperate to avoid his father's fate, haunted by August Heinrich's demons. He began to closely regulate his family's finances, curating thousands of bits of information in bound journals. About ten years earlier the scientist began charting his wife's domestic expenses by collecting reams of paper filled with prices of groceries, insurance, clothing, literature, and anything else that had cost him money. The type of data was not alarming, but the amount of information was incredible. Storing it all made him feel more in control of his money, even when he was not.

In 1915, the criminalist defaulted on his mortgage for his home in Tacoma, and the legal process had stretched over three years. When the Heinrichs relocated to Boulder, Kaiser began collecting their mail in Tacoma. In 1918, the librarian sent an urgent telegram to Oscar about threats from the bank.

"Bankers Trust Co has foreclosed mortgage and secured judgment amounting to over eleven hundred dollars," read Kaiser's telegram. "Property will be sold unless you adjust matter."

Oscar sent a payment and regained ownership of the house, but money struggles continued to plague him. And in 1921, despite his mounting debt, Oscar considered opening an additional lab, a smaller one in San Francisco for clients in that city. It might have seemed like a horrible decision, but Oscar was determined to remain a successful (and competitive) forensic scientist. He had to maintain adequate facilities—the willingness to grow his business was a requirement, but it was also a dangerous risk.

His father's sudden death solidified the sixteen-year-old boy's fate—life would be a constant struggle. His father's death roused Oscar's determination to skirt poverty for good. He was determined to offer his two young sons a happier life—a stronger, more stable upbringing than his own, with plenty of toys, food, and joy.

Now in 1921, Oscar Heinrich was ready for his first case to make "E. O. Heinrich" the star of international headlines. August Vollmer would soon label him "the foremost scientific investigator of crime in the United States." It was quite a heavy burden for a man already carrying a lifetime of angst and self-doubt. Yet Oscar buried his fears and moved forward toward his first big investigation, one that would put him on the front pages for the first—and definitely not the last—time.

3.

Heathen: The Case of the Baker's Handwriting, Part I

There can be no question as to the authorship. See how the irrepressible Greek e *will break out, and see the twirl of the final* s. *They are undoubtedly by the same person.*

—Arthur Conan Doyle, *The Sign of Four,* 1890

The fog swallowed tiny Colma, California, in a heavy, gray mist that moistened ragged overalls and settled on the granite headstones that lined Holy Cross Catholic Cemetery. It was August 2, 1921—a chilly Tuesday night that encouraged people to spark small fires inside brick hearths.

Colma surely must have been haunted. For decades the town had thrived off of death, requiring the deceased to support the living. Back in 1849, hundreds of thousands of people had swarmed nearby San Francisco, prospectors mining for gold but carrying along deadly diseases. Fortunes were made, but not without a significant number of deaths. There were more bodies than the city could handle.

Within fifty years, San Francisco had banned all burials—the land was too valuable to squander on cemeteries. In 1914, San Francisco began evicting bodies from its graves—thousands of corpses were exhumed and shoved onto macabre streetcars headed south. The "cemeteries line" originated at Mission Street and stopped in Colma, about ten miles away. The whole process cost families about $10 per loved one; otherwise the bodies were relegated to mass graves.

In the years that followed, the electric streetcars emerged daily from the fog like props from a gothic novel: dark-colored carriages resembling caskets with curtains draping the windows. Inside, the coffins slid atop woven rugs while grieving families rested nearby on comfortable wicker chairs during the hour-long trip. "Funeral Car" was emblazoned in gold on the fronts of the cars, an unambiguous signal to their purpose when they arrived in the town.

Colma occupied just two square miles of land, but much of it was covered with graves. Funeral processions were daily rituals, and the interred bodies were some of the most famous in America, legends like outlaw Wyatt Earp and denim magnate Levi Strauss.

Most of the living people in Colma were employed in some type of cemetery work, and within twenty years the town of a few hundred residents had inherited 150,000 bodies. In 1924, it was designated a necropolis, one giant cemetery dubbed the "City of Souls." The constant presence of mourning families compelled the churches to comfort the bereaved, so priests in tiny towns like Colma were revered in 1921, seemingly untouchable by evil.

Father Patrick Heslin, the priest for Holy Angels Catholic Church on San Pedro Road, sat quietly in his study as the fog swirled around the street just feet from his front door. His housekeeper of seven years, Marie Wendel, flitted around the kitchen of their parish house, a large building connected to the church. The fog in the waning twilight muted the car headlights of the few drivers who braved the road.

At fifty-eight years old, Father Heslin was handsome, tall, and husky, with thinning dark brown hair and an Irish brogue. A native of County Longford, the priest was a kind but authoritative leader behind the pulpit on Sundays. Father Heslin and his housekeeper had been in Colma just ten days, but while at his previous church in Turlock, about one hundred miles away, Heslin had provided guidance for the devout, counsel for couples in turmoil, and education for young believers. The priest hoped to do the same in his new parish.

Across the street from Father Heslin's home, a motorcar's engine hummed and then switched off. Footsteps crossed the road. The home's doorbell rang, followed by a loud knocking and a frantic pounding. Father Heslin listened as the housekeeper unlatched the lock and swung open the door.

"There's a man who wants you to minister to a dying friend," the housekeeper told the priest.

Marie Wendel was clearly unsettled. The visitor was fidgety, just moments from becoming unhinged. He was dressed bizarrely for night-time travels, and he had firmly refused to come inside despite her insistence.

"I'm in a hurry," he anxiously explained in a heavy accent.

Wendel felt unnerved, while Father Heslin appeared puzzled. He walked toward the front of the house, curious to hear from a stranger who might venture into a thick fog for a good friend. As he approached the entryway, the priest's eyes widened.

The man was almost smothered by a huge slouch hat. He wore a large, heavy coat with the collar turned up, shielding his face. He seemed to have tanned skin—a signal to the housekeeper that he wasn't American. He wore dark-colored goggles, a common accessory for drivers tooling around in open-air 1920s touring cars, but during nightfall and in this gloom the eyewear seemed ludicrous. A nosy neighbor eyed the stranger's car parked in front of her house as she stepped outside. She strained to hear the conversation.

Wendel relocated her housework to another room to offer them privacy. The visitor explained that his friend had suffered from tuberculosis. He was doomed, the stranger believed. He desperately needed his last rites read to prepare his soul for death. Father Heslin nodded—he agreed to pray with the dying man and allow him to confess his sins.

Tuberculosis was the leading cause of death in the United States in the nineteenth and early twentieth centuries. In fact, the American

Lung Association was created in 1904 specifically in response to the deadly disease. In the 1920s, tuberculosis, also called "consumption," was a death sentence, one that afflicted mostly poverty-stricken city dwellers with fatigue, night sweats, and violent coughs before finally killing them.

Father Heslin knew it would likely be a perilous trip because tuberculosis was extremely contagious—even some doctors refused to treat sufferers—but the priest's beliefs were grounded in self-sacrifice. Wendel returned to the entryway and quietly asked Father Heslin if he was responding to someone who was dying. *Yes,* he replied.

"I will be back as soon as possible," the priest assured her.

Wendel latched the front door as the peculiar man settled into the driver's seat of his small Ford touring car and pushed the button to start the engine. She peeked from behind the heavy curtains of the front window when he swung the car around toward the nearby highway and parked again to wait. The housekeeper watched the priest quickly open the passenger door. Within seconds Father Patrick Heslin had vanished.

National magazines had bolstered the popular notions of the twenties as a decade of glamour, jazz, bathtub gin, and swinging speakeasies, but for many Americans the new decade did not roar like one of Jay Gatsby's parties.

The conclusion of World War I in 1918 did not revitalize the economy as the government had promised. Soldiers returned home traumatized, angry, and often with little hope of finding jobs. Most people struggled to provide for their families—the unemployment rate had more than doubled to nearly 20 percent in the twelve months following the end of the war. The go-go, prosperous twenties wouldn't really begin until 1923; in the meantime, widespread poverty swept across the country, unalleviated by Warren Harding's recent election to the White House.

The year 1921 also marked the first anniversary of Prohibition, another reason for despair among many Americans. Since the law had been enacted, the national crime rate had increased by almost 25 percent; police were cracking down on illegal alcohol consumption, and criminals crowded the prisons. Police were preoccupied with enforcing strict anti-drinking laws, so the Bureau of Prohibition "deputized" Ku Klux Klan members, who used their power to harass poor immigrants, along with black Americans.

Despite the ban on alcohol, drinking had become an epidemic, and arrests for drunken driving rose by 80 percent. Other criminal enterprises flourished as well. In the 1920s, loosely formed gangs forged an organized crime syndicate that produced a dramatic increase in criminal activity. American mafia leaders like Al Capone thrived as bootleggers, particularly in large cities like New York and Chicago. The Newton Boys, the most successful train bandits and bank robbers in history, hit six trains in the 1920s, though bandits were more frequently met with armed guards. All across the US, detectives struggled with increasingly difficult investigations, and in August 1921, the mystery of the missing priest simply bedeviled the brightest minds in law enforcement.

Marie Wendel fretted when Father Heslin didn't return by midnight that foggy night. She rambled around the unfamiliar two-story house, trying to stay distracted with sewing projects. She finally reasoned that last rites could be a long process if the victim had been reluctant to die. Wendel fell asleep, hopeful that she might hear the church's lovely bell toll in the morning—ringing it was one of Father Heslin's many duties. But the next day, when Wendel found the belfry silent and the priest's bedroom empty, she phoned the archbishop's office in San Francisco.

"He left with a small, dark man, probably a foreigner," she told his clerk.

Archbishop Edward Hanna dismissed Wendel's concerns; a priest not notifying his housekeeper of his schedule did not seem cause to raise an alarm. But soon his clerk shuffled back into his office, carrying a special delivery note with a two-cent stamp. Hanna quickly opened the letter and carefully read it over. Now it was the archbishop's moment to feel unnerved.

Oscar Heinrich adjusted his glasses as his eyes followed the curve of the letters scrawled on the paper in his hand. He stood silently in the U.S. Postal Service's San Francisco office, gently holding a letter. It was just days after Father Heslin had disappeared, and the criminalist could sense the overwhelming urgency from the investigators hovering nearby. Sometimes detectives were valuable, like Berkeley's brilliant chief of police, August Vollmer. But much of the time, cops (who loved to call him "Doc") just addled his process. Oscar hadn't taken the time to assess this group just yet.

The forty-year-old steeled himself for their unreasonable, emotional reactions; the police in San Francisco were sensitive to a potential backlash from an alarmed public. Only a devil would hurt a Catholic priest, thought Oscar.

He suspected that hiring him had not been the investigators' idea. The archbishop summoned him because detectives were baffled by the strange, almost incoherent letter that arrived at the parish office shortly after the priest's disappearance.

"Had to HITT him four times and he unconscious from pressure on brain," it read. "So better hurry and no fooling. TONIGHT at 9 clock."

It was a ransom note, more than six hundred words long, both handwritten and typed—as if two people were quarreling over control of the letter. The author demanded $6,500, which seemed like an odd amount to Oscar. Most kidnappers tended to round up or down. The typed sentences were well composed, while the handwritten demands were misspelled and confusing.

Act with caution, for I have Father _____ (of Colma) in a bootleg cellar, where a lighted candle is left burning when I leave, and at the bottom of the candle are all the chemicals necessary to generate enough poison gas to kill a dozen men.

Oscar pulled the note to his face as he rubbed the paper between his fingers—nothing distinct about its composition. The kidnapper's claims were horrifying if they were credible or creative if they proved to be fiction. Oscar looked at the space left after "Father." Clearly Patrick Heslin had not been the specific target, but this crime was certainly premeditated. The kidnapper was likely a very clever man, or insane. Or both.

The note warned that Father Heslin would die if the police were called—he would burn to death, shackled in that bootleg cellar. The kidnapper's rambling note was menacing, and his military background, he claimed, had prepared him to be an assassin.

"I had charge of a machine gun in the Argonne, and poured thousands of bullets into struggling men, and killing men is not novelty to me," the letter read.

The remainder of the note read as if penned by a villain in a dime-store detective novel. The kidnapper ordered the archbishop to quickly prepare the money in a sealed package because he would soon receive driving instructions.

"Get out with the money, leave the car and follow the string that is attached to the white strip [on the road] until you come to the end of the string," read the letter. "Then put down the package and go back to town."

Oscar stood nearby as the archbishop explained how he had waited faithfully for driving instructions that had never arrived. In the meantime, San Francisco's chief of police authorized an aerial search for automobile wreckage along the foggy coastline. Hundreds of volunteers led by bloodhounds fanned out along the route toward Salada Beach and Pedro Mountain in the direction where the stranger's car had headed. Investigators peered over cliffs hanging above the ocean as a tantalizing mystery in California quickly became a national spectacle.

Newspapers across the country immediately caused confusion by printing baseless conspiracy theories. "Was Priest Kidnaped to Wed Pair?" asked the *Oakland Tribune*. San Francisco detectives had refused to release many details of the case, but one officer did offer the press a small clue: "Someone needed a priest and needed him badly."

"This significant statement," wrote the reporter, "could only mean the performance of a marriage, perhaps under duress."

Investigators might have been amused by the ridiculous speculation, but protecting genuine clues was challenging because the town of Colma, a speck on the state map, was now infested with reporters. Newspapermen relentlessly harassed Marie Wendel, the priest's housekeeper, as journalists accused her of being an accomplice in the kidnapping.

"I have nothing further to say," Wendel screamed to a crowd of journalists. "I have been widely quoted with statements which I have never made at any time."

"Why did you not report the disappearance of Father Heslin's automobile?" snapped a reporter.

"The automobile was not missing!" she replied.

Father Heslin simply had it sent to a local repair shop. But still, misinformation and rumors about the case proliferated. Three different neighbors agreed that the car's driver "did not appear to be an American," and now every newspaper claimed that a foreigner had snatched the beloved priest. The absence of information sent reporters into a frenzy, stoking the already xenophobic atmosphere in the predominately white town.

Detectives were stymied, and the only lead they had was the peculiar ransom note. They hired three renowned document experts, analysts who would compare the printing or handwriting on documents to either link or eliminate someone as the source. They each claimed to be able to reveal alterations, additions, or deletions of characters; they used chemical analysis to connect the ink on a letter to a specific pen.

One expert was Oscar Heinrich, whose lab was already gaining some fame in the discipline of questioned documents by 1921. Another expert was a man who would become Oscar's longtime nemesis, a painful thorn in the investigator's side. The case of the missing priest would spark a horrible, sometimes shameful rivalry that concluded only when one of them suddenly died almost fifteen years later.

But what clues were these men looking for in the ransom letter? A few things. There were two disciplines in the field of handwriting detection: handwriting analysis and graphology. Even today they're often lumped together, but they're actually quite distinct.

Handwriting analysis has a long and storied history. It was developed as early as the third century, when judges in Roman times compared signatures and other lettering in documents to determine forgeries. While much of the analysis was based on mere guesswork, by the late nineteenth century, formal educational classes in handwriting analysis were starting to be offered to the public. But the "experts" who claimed to understand handwriting analysis were generally self-trained and self-certified, and the quality of their analysis varied greatly.

By the early twentieth century, experts in forensic handwriting analysis were frequently called to testify in the cases of forged documents. Just a few months before Father Heslin's disappearance, Oscar had worked with another forensic document expert on a forged will case in Montana, a dispute over an estate worth more than $10 million. They were able to prove that the will's signature was a forgery, a celebrated but modest legal victory. Slowly, handwriting analysis was gaining respectability and acceptance in the legal world. And these handwriting analysts weren't just relegated to matters of wills and estates, either. More and more, as the twentieth century progressed, they were starting to find their way into criminal courts. Lawyers and juries became enamored of the idea that certain pieces of evidence, such as shoe impressions, fingerprints, bullets, or bite marks, were unique, individualized. If handwriting analysis was admissible in a civil investigation, why shouldn't investigators start using it as a tool in more serious crimes, too?

The idea behind handwriting analysis was that only one person could make those exact letters in that *precise* way; therefore, a signature or a longer written passage would identify an individual as clearly as a fingerprint might. Today we know this isn't true—handwriting style can vary widely depending on the author's environment, his writing tools, his age, or even his mood. Our handwriting is too unpredictable, so the forensic technique is mostly unreliable.

A 2009 landmark study by the National Academy of Sciences, which investigated and evaluated an array of forensic science techniques, determined that some handwriting samples aren't unique enough to determine fabrication. "Some cases of forgery are characterized by signatures with too little variability, and are thus inconsistent with the fact that we all have intrapersonal variability in our writing," read the report. In other words, since one's signature can vary by day, depending on a whole variety of factors, signing a name to a document isn't equivalent to, say, leaving a fingerprint behind.

But the committee agreed that there was some scientific value to handwriting analysis, especially in the case of comparing clear, well-constructed words with specific attributes. Longer passages of handwritten text could show repeating characteristics, and there were clues to be found in comparing lengthier samples of writing. But it was often inclusive. Bottom line: handwriting analysis can sometimes be used to support an investigation, but never to close one. Yet back in the early 1920s, expert testimony involving handwriting analysis was starting to be deemed admissible in criminal courts—Oscar's business as a questioned-documents expert was flourishing.

Alongside the study of handwriting analysis was another much less reputable discipline: graphology. Some said that graphology was to handwriting analysis what astrology was to astronomy—more art than science. Whereas handwriting analysis dealt in matching characteristics of two pieces of text to each other, graphology, its practitioners insisted, would illuminate the *personality* of an author—the psychology behind his handwriting strokes, like rudimentary criminal profiling. Graphologists claimed to be able to predict the suspect's state of mind at the time of writing a note. For example, graphology experts claimed that a person who had nothing to hide would likely leave a small gap at the top of the letter *O*. An author who closed the oval loop in his lowercase *g* might lack confidence in sexually intimate situations.

Graphologists weren't particularly well respected, and forensic graphology (the modern term) has been considered a pseudoscience for more than one hundred years. Oscar Heinrich was not a graphologist, to be sure, and he even ridiculed the practice in a handwritten letter to his best friend, John Boynton Kaiser. In it, Oscar performed a bit of self-analysis based on his own pen strokes.

"Notice how mine goes over the top . . . now, up as in ambitious flights of optimism," Oscar quipped. "Now down, the picture of pessimism. Don't be alarmed. I have merely been writing over my wife's billowy desk blotter."

Oscar knew that graphology was nonsense and even dangerous because investigators might ignore relevant clues if they depended too heavily on an inaccurate psychological profile of a suspect. But the public and the police demanded answers, and in the case of Father Heslin, the graphologists' insights offered hope. The police brought in two graphologists to examine the ransom note, and Oscar firmly disliked both of them. Carl Eisenschimmel was a German whose work Oscar found to be careless. The other was Chauncey McGovern, a local expert who Oscar considered not just dotty but dangerous. The forty-eight-year-old was destined to become a scourge on Oscar's professional reputation, but for now he was simply an irritant.

Oscar was dismayed that the police required anyone else's opinion, but as usual, he was forced to muster some diplomacy. He was still early on in his career, after all, so good customer service was imperative. Oscar shook hands with the other men and stood to the side.

Eisenschimmel and McGovern concluded that the mysterious author of the note was likely an ex-soldier, a military typist. A boldly specific claim, to be sure . . . except that the writer himself had mentioned his time in the army. Hardly groundbreaking insight, thought Oscar.

The experts went further: based on the shape of his letters and the generally schizophrenic nature of the note, Eisenschimmel and McGovern were certain that the kidnapper was insane.

"The writer is demented," concluded McGovern with confidence. "The style of the printed 'H' and 'A' is that of a deranged person."

Carl Eisenschimmel, the other expert, quickly agreed: "The block printing at the bottom of the ransom letter is a script often used by demented persons."

Oscar sighed. Those conclusions weren't very helpful—they weren't specific enough. They were wasting time, as usual. Eisenschimmel was an old curmudgeon, a disagreeable egotist. And despite a few high-profile gaffes in his career (his ego had led him astray a few times, often disastrously and in open court), the seventy-five-year-old German was still held in esteem by jurors and investigators he testified in front of. With his full white mustache, immaculate three-piece suits, and thick accent, Eisenschimmel oozed professional gravitas, something that Oscar clearly envied. He had testified as an expert witness on forged documents for decades, since Oscar was no more than a teenager. And on this investigation, the German criminalist was tapped as the lead expert, while Oscar was relegated to second, even third chair. Oscar was miffed over being forced to offer deference to the elder expert, an insult for a sensitive genius whose own ego was building with each successful case.

"I am expected to do all of the heavy work of preparation," he complained bitterly to John Boynton Kaiser, "wheedle the old man into the belief that everything is being done according to his suggestions, and probably let him take the credit for the case."

It was infuriating for a man of his rigorous methods to play second fiddle to an expert whose approaches and science were so suspect. Eisenschimmel was inept. But Oscar labeled the other expert, Chauncey

McGovern, as worse: smug, pompous, and even derelict. He was convinced McGovern was a quack, and his unreliable methodology was a noose around the necks of legitimate handwriting analysts. McGovern, too, had some unsavory aspects to his résumé, and Oscar worried that his uneven (even unethical) record might discredit the whole field of handwriting forensics for good.

Almost twenty years earlier, attorneys representing the American government had arrested McGovern for perjury, accusing him of exaggerating his credentials during his expert testimony in a different case. The United States Supreme Court eventually cleared him of those charges, but his reputation was still tarnished. Yet during the nascent era of forensics—where just about anyone could claim to be an expert in any discipline—McGovern was considered a safe bet for prosecutors because of his composure on the stand. These two "experts" were Oscar's primary competitors: a dunce and a liar, according to him.

Throughout Oscar's long career Chauncey McGovern would solidify his role as "chief antagonist." But he was just one of many forensic experts who tried to discredit Oscar on the stand during public dustups, and the criminalist would do the same to them. Their vicious duels in court would embarrass the field and confuse jurors. But for now, Oscar Heinrich and Chauncey McGovern were both being paid by the prosecutor to identify a kidnapper using the man's own hand. U.S. Postal inspectors handed Oscar the ransom letter. He glanced over it one last time.

"There's at least one thing I can tell you about this right now," he told the investigators. "Whoever wrote that letter revealed his trade."

"What do you mean?" one asked.

Oscar peered through his spectacles.

"The writer of that letter is a baker," Oscar replied.

The investigators snickered. "Just what do you mean by that?"

"This is cake baker's lettering," Oscar explained, "the kind that all master bakers learn."

The postal inspectors seemed to stop smiling.

"Examine closely the concave uprights of the *A* and the *H*," he replied, "the down curving crosses of the *T*, the square bottom of the *U*. That's the style bakers use in writing on cakes. Look at the frosted lettering on your next birthday cake."

The investigators were incredulous, but Oscar had not been hired to placate skeptical cops—he was ordered to remain impartial, to follow the evidence.

"But there are a good many bakers in town," the detectives argued.

Oscar smirked—just as he suspected: the police were likely to get in his way. In early criminal investigations, most profiles were completed after the suspect was in custody, usually to establish sanity. Oscar wanted to discover the identity of the suspect *before* he was caught, a new approach. He focused on clues that revealed habits, an early form of criminal profiling—like nothing seen before in America.

Offender profiling had been used sporadically in history, beginning in the Middle Ages with "experts" trying to identify heretics. The first well-known psychological profiling in a criminal case involved inspectors at Scotland Yard in 1888, who hoped to glean information about Jack the Ripper from his letters and the condition in which he left the bodies. Most investigators in America wouldn't use profiling until the mid-1950s, and it wasn't until the 1970s that the FBI would establish its Behavioral Science Unit. But Oscar was a man ahead of his time; from the beginning of his career, he believed he could reconstruct a crime scene by visualizing the habits and actions of the criminal.

"I do this by using the evidence that the criminal leaves behind," he explained.

McGovern's and Eisenschimmel's claims of determining the sanity of the author were ludicrous, according to Oscar. All criminals have habits, he argued, that were inescapable, hidden even from the killer. And those habits would give away his identity. A baker who committed murder remained a baker, which was why the ransom note was written with unusual lettering. His actions were habitual, almost uncontrollable, and they would help establish a profile. Oscar firmly believed that investigators should study the evidence to solve the criminal's identity before settling on a motive, but the proof would be in his results. As detectives continued their search for a devil somewhere in the hills near Colma, California, Oscar made them one promise.

"Your man is a baker."

The case was a nightmare for the archbishop in San Francisco, who prayed each night that Father Heslin would be found safe. But it was already August 10, eight days after the priest had gone missing. A new ransom note, slipped under the door of the archbishop's residence overnight, promised that the priest was still alive—for now. And it was even more peculiar than the first—a rambling narrative loaded with mysterious clues that seemed designed to lead investigators into more blind alleys.

"Fate has made me do this," wrote the author. "Sickness, misery, has compelled my action. I must have money."

The kidnapper issued his new terms, which included a $15,000 ransom, more than double the initial amount.

"Father Heslin is safe and says for you to help him," the note read. "You will hear from me very soon."

The sheriff called back the two other handwriting experts, McGovern and Eisenschimmel. Oscar wasn't consulted; perhaps his earlier analysis wasn't what the investigators wanted to hear. The pair of experts enlarged the note, illuminated it with a lamp, and determined once again that the author was certainly a lunatic and a coward.

"The handwriting shows the writer to be a jellyfish," was McGovern's official report.

The expert's conclusion was yet another useless observation—there were certainly loads of insane criminals in Northern California. *Pointless,* scoffed Oscar.

In the meantime the unfounded rumor of a third letter, secretly delivered by a detective to police headquarters, piqued the interest of a cub reporter with the *San Francisco Examiner* named George Lynn. With few fresh leads and a slew of eager readers waiting for new details, Lynn's editor dispatched the journalist to the archbishop's home in the hopes of a new scoop. Lynn was about to become the envy of every newspaperman in the state.

He hailed an unlicensed cab, called a jitney in San Francisco, and arrived late that afternoon at the archbishop's mansion on Fulton Street at the northwest corner of Alamo Square Park. As George Lynn rang the bell, he turned to see someone else climbing the wide stone steps. Lynn took a step back and surveyed the man's clothing, a strange tropical ensemble that seemed unsuitable for a metropolis like San Francisco. His dapper cream-colored outfit made from a blend of mohair and cotton, complete with a straw hat, would have been perfect for the hot, sultry streets of Palm Beach, but he looked out of place in San Francisco. Lynn eyed the man with curiosity.

The lanky stranger stood quietly near Lynn as the archbishop's door opened. The archbishop's assistant eyed the white-suited stranger warily and then nodded at the reporter. Was His Holiness at home? the newspaperman asked. The archbishop was finishing dinner, explained the assistant, but Lynn was welcome to wait inside. The man in the tropical outfit slid through the door as they both walked into the parlor.

"I don't know this man," Lynn said quickly, hoping the archbishop's assistant would eject the stranger. "He just walked in with me. I'm not responsible for him."

"I came to see the archbishop, too," the man snapped. "I think I know where the missing priest is."

William A. Hightower was the stranger's name, and as he talked, Lynn silently remembered the witnesses' description of Father Heslin's kidnapper: "a small, dark man, probably a foreigner." He looked over

Hightower and decided that he could not possibly be that same man. Hightower was tall at about five feet ten, thin with medium-toned skin, and he was almost bald. He hailed from Texas and was raised on a cotton plantation, where he'd been a worker in the fields since he was young. He had a southern drawl, not a foreign accent. Lynn grew more relaxed as he listened to the man's story. Hightower was certainly excitable, but he wasn't the kidnapper. Lynn eyed the man's light-colored suit again as Hightower smiled. "I've just arrived from Salt Lake City," he explained, "and it was much warmer there."

When the archbishop finally entered the parlor, he turned to Lynn and declared there was no third note and no scoop. The reporter nodded and introduced Hightower, and then both men listened as he narrated his wild story. The tale starred a "nightlife girl" named Dolly Mason and a dangerous stranger, a bootlegger with a foreign accent who bragged of burying mysterious cargo at Salada Beach, about twenty miles south of San Francisco. Hightower also added another distressing detail—Dolly said the foreigner seemed to despise Catholics.

The archbishop was skeptical of the man in the strange clothing, and he had no use for nonsense. Investigators had been chasing false leads for more than a week, and this tale of showgirls and bootleggers seemed to have no bearing on the search for Father Heslin. The archbishop refused to believe Father Heslin was dead, but he also suspected the man standing in his parlor was truly daft. However, there weren't a lot of other clues coming in, despite massive searches, hundreds of volunteers, and even a sizable reward for the missing priest: $8,000.

"If you will return tomorrow about ten in the morning," the archbishop assured him, "I'll send a couple of investigators with you."

George Lynn's eyes widened. This guy Hightower might be a nutter, but Lynn knew a good story when he heard it. As the pair left the archbishop's mansion, Lynn raised his arm, hailed another jitney, and swung open the door, inviting Hightower to join him for a short trip. The reporter ordered the driver to go directly to the offices of the *San Francisco Examiner.*

"He says he might be able to tell us where Father Heslin's body is," Lynn explained to his city editor, William Hines.

The pair was quickly swept into a private room, where Hightower recounted his bizarre conversation with Dolly Mason a few days earlier. She had believed that the stranger was concealing a cache of bootleg liquor, a scheme that wasn't uncommon in the 1920s.

Even during Prohibition, there were many opportunities to drink, and not just in speakeasies. Some people simply asked their doctors for a prescription, as pharmacists actually distilled medicinal whiskey for patients. In fact, that had been one of Oscar's duties at his own pharmacy when he was a teenager in Tacoma. Medicinal alcohol was such a profitable business that it allowed Walgreens pharmacies to expand from around twenty stores to more than five hundred during the 1920s. But if you weren't lucky or rich enough to get a prescription, there were still many entrepreneurs willing to supply alcohol to you—illegal liquor was valuable merchandise, and the idea that Hightower's friend Dolly Mason knew a less-than-savory character who was determined to protect his stash on Salada Beach was a believable prospect. But what did it have to do with Father Heslin?

Hightower was relishing the attention, dragging out the story for dramatic effect.

"So Sunday I took a run down there to the beach," Hightower said slowly. "I remembered there was a billboard down there—one of those that shows a miner frying flapjacks over a campfire."

George Lynn suddenly squirmed in his chair. *Flapjacks.* His heart raced. The pancakes reminded him of a detail offered by one of those handwriting experts mentioned in his stories, that outlandish clue from criminalist Oscar Heinrich in Berkeley. *Something about cooking?* . . . The journalist wanted to test Hightower, but he had to be clever.

"By the way," said Lynn, abruptly cutting in, "I forgot to ask you, but just what do you do for a living?"

Hightower glared at him, suddenly peeved by the reporter's rudeness. George Lynn was certain of how the man would reply, but he wanted to hear it said aloud.

"What difference does that make?" Hightower asked sharply. "If you must know, I'm a master baker."

The headlights from the police cars illuminated a ghostly image on the giant billboard advertising Albers Mill flapjack flour, just off the side of the Ocean Shore Highway above Salada Beach. The sign showed a drawing of a weathered miner with a handlebar mustache who was tossing a flapjack over a campfire in the desert, just as Hightower had described. The sea mist from the warm Pacific Ocean floated around the sign, almost in waves; the black road felt slick underneath the shoes of San Francisco's police chief, Daniel O'Brien, as he stood at the edge of the cliff. It was around eleven as the group of seven men peered below while the fog drifted up from the sand dunes and the waves slammed against rocks on the shore. It was an eerie mission—a gang of ghouls hunting for a body in the fog.

A few weeks earlier a production company in Germany began shooting what would become one of the greatest horror films ever made, *Nosferatu: A Symphony of Horror.* In the film, a gaunt, pointy-eared creature named Count Orlok stalked his victims inside his castle in Transylvania—the first vampire depicted on the silver screen. The horrible anticipation of Orlok's attacks was indelible for viewers. That nervousness and fervor would be echoed for the group of men standing above the Pacific Ocean as they allowed a suspected killer to lead them down the side of a cliff—a gorgeously gothic scene.

William Hightower, George Lynn, and a group of policemen stumbled over rocks down a narrow path, which led them down the side of a sand cliff. It was wet and windy as they lit matches for guidance through the fog. They climbed lower along the cliffs—the sounds of the ocean's surges were deafening as the men hauled along shovels and picks. Someone would later visit a nearby artichoke farm in search of lanterns as a cameraman for the *Examiner* snapped photos from a safe distance. It was a dreadful night to dig for a corpse.

Hours earlier Hightower had informed George Lynn (and later the police) that he thought he had found a burial spot for *something*.

"Maybe it's that missing priest," he told police. "That's what I've been thinking. The ground's loose too. Seems to me that's worth looking into, don't you agree?"

As the group slowly traversed the rocky trail in the dark, it was hard for investigators to settle on what they hoped to discover. The murder of a beloved religious leader was blasphemy, an act that could only be linked to a godless monster. But finding Father Heslin's body would at least provide his loved ones with closure. Neither outcome would assuage Americans as the story continued to make national news.

Suddenly, Hightower hopped down to a low edge on the side of the cliff. As the others slowly followed, he dropped to his knees and shoved his hands into the sand, yanking up a piece of black cloth, a marker he said he had left just a few days earlier.

"There it is," Hightower yelled. "That's the spot."

George Lynn and an officer, armed with shovels, began slowly digging like nineteenth-century grave robbers praying for a big payday. Hightower gripped his own shovel and began wildly flinging sand over his shoulder. Silvio Landini, the constable for the town of Colma, stepped back.

"If the body is in here, you ought to be a little easy with that shovel," he warned. "You might strike the face and mar it and we don't want to do that."

Hightower paused and looked up. "Don't worry," he replied. "I'm digging at the feet."

The other men stopped and glanced at one another. The police chief stepped cautiously toward Hightower as the baker continued to hurl sand behind him. Reporter George Lynn stopped suddenly—his shovel struck cloth buried just beneath the surface of the sand. He began to slowly lift it up.

"A hand," the police chief yelled.

Within minutes they exhumed a heavy corpse wrapped in damp, black clothing. Hightower, now exhausted, proudly stood at the feet. There was no doubt that the dead man was Father Patrick Heslin. His sacred vestment, the one he had carried from his home eight days earlier, was still draped around his neck. His skull was bloodied, bashed in and partially gone. As the men waited for the coroner, Hightower stared at the body and quietly mused.

"Human life is a funny thing," he whispered to the chief, who glared back.

"Let's go!" Chief O'Brien bellowed before dragging Hightower up the cliff to his squad car.

George Lynn peered down at the priest's body as it lay among the patches of green devil grass. The dim light of a lantern revealed a small picture of Christ.

4.

Pioneer: The Case of the Baker's Handwriting, Part II

How often have I said that when you have excluded the impossible whatever remains, however improbable, must be the truth.

—Arthur Conan Doyle,
The Sign of Four, 1890

"Get Ready." That was the message from *San Francisco Examiner* city editor William Hines to his publisher. When George Lynn returned late that night from Salada Beach, the seasoned editor sensed that the horrible murder of a priest would become a sensational, twisted news story—an exclusive for his paper. San Francisco's chief of police spent the night at the newspaper's headquarters while police deputies protected the building from spies hired by other papers. The police had struck a deal with the *Examiner*: George Lynn's delivery of Hightower to police in exchange for an exclusive story. The presses rumbled throughout the night, and the next day the *San Francisco Examiner* sold thousands of copies.

Three days later, after an immense amount of questioning in jail, police formally arrested William Hightower for Father Heslin's murder, and soon the mechanism of criminal law jurisprudence suddenly started to move. Investigators didn't believe Hightower's harebrained story about a foreign bootlegger who confessed to a tart named Dolly Mason, but still they conducted a massive search for the woman. They would find her, Hightower insisted, and he would certainly be exonerated.

"You've got a funny way of showing your gratitude when I solved your case for you," Hightower complained to detectives.

The coroner's report on Father Heslin's death was distressing. The priest had been beaten to death with a blunt object and then shot in the back of the head with a .45-caliber gun as the killer stood over him.

Police gathered spent shells, white cord, and blood-stained wood at the gravesite on the beach. These pieces were part of a very convoluted jigsaw puzzle, a mystery that San Francisco police couldn't quite sort out. Detectives quickly realized that Father Heslin's murder would not be solved by handwriting experts with limited forensic experience. The captain ordered his officers to hand-deliver the blood-stained wood and string to Oscar Heinrich's lab in Berkeley.

Detectives searched Hightower's cheap lodging house room in the Mission district of San Francisco and collected more evidence: pieces of bloody burlap, a rifle, and newspaper clippings that mentioned a reward. There was also something startling, even to experienced cops—a homemade weapon that police dubbed an "infernal machine." It had ten short lengths of pipe, all stuck into a wooden frame, and each pipe held a shotgun shell. Hightower had designed it so that he could pull a long string and shoot all ten cartridges, an archaic machine gun.

"Follow the string," the first ransom note had ordered. Detectives suspected a link between the machine and the kidnapper's veiled instructions. An officer pulled out a heavy canvas tent hidden under Hightower's bed with the word TUBERCULOSIS printed in large block letters. There were more .45-caliber shells, a machine gun, a gas mask, and some poetry apparently composed by Hightower.

Detectives searched for the contours of a broader scheme—a clear motive. But what they had gathered was a confusing collection of material from a man who seemed to be unmoored from reality. Yet investigators still had doubts because their suspect didn't fit the description of the kidnapper, "the small, dark foreigner." Hightower continued to deny the murder from his San Francisco jail cell and declared that Dolly Mason would provide clarity, if police could only find her.

Despite the new discoveries, the case against William Hightower was still largely circumstantial with few forensic clues; prosecutors admitted that an assortment of curious items found inside the dingy room of a flophouse would not secure a murder conviction.

On August 13, eleven days after Father Heslin disappeared, Archbishop Edward Hanna blessed and eulogized the priest at St. Mary's Cathedral in San Francisco. Bells tolled as thousands of people stood outside the church while more than four hundred members of the clergy watched the service inside.

"Father Heslin has made the supreme sacrifice," Archbishop Hanna told the crowd. "He has shown the greatest love it is possible for man to give. His end is one that any priest might ask for."

While the Catholic community in California honored Father Heslin, Oscar Heinrich examined evidence, determined to solve his murder. The police soon summoned him back to San Francisco for an update—once again, Oscar reveled in their anxious stares. He would be their savior, he was sure of it.

"Somewhere there is a clue to the method followed by the man who committed this crime," Oscar told detectives. "What did Hightower have on his person when he was arrested?"

An investigator opened a manila folder. As a small pocketknife with a dark handle slid out onto the table, the detective assured Oscar that there was nothing to glean from it because police officers had already examined it thoroughly. The criminalist knew better.

Oscar returned to his lab in Berkeley, settled into his wood chair, and slipped the hilt of the knife under his microscope. He adjusted the oculars, narrowed his eyes, and spotted them—a few grains of sand. Oscar increased the device's magnification, brightened the light, and placed his camera over the eyepiece, snapping a photo. He slid over a different microscope sitting on his long wooden table. Oscar glanced at his notes about the grains of sand that he had discovered adhered to the hat found inside William Hightower's room. He examined the steel blade of the knife.

"There is a small patch at the foot of the large blade," he wrote in his notes, "which corresponds in appearance and size of grain with the sand from the hat."

Hightower's hat contained sand that likely came from Father Heslin's gravesite—a solid piece of circumstantial evidence.

"Now I'm going to work on the other things," Oscar told police, "the tent in his room, the lumber from the graves, everything."

He unraveled the tent with TUBERCULOSIS printed on the side and looked closely at the block lettering. Investigators suspected that Hightower had used the ruse to discourage anyone from looking inside the tent.

"In form this writing corresponds with the writing on the ransom note," Oscar said. "But the writing movement however is so different and the material upon which it is written is so different from the ransom note that a further comparison seemed to me inadvisable."

Oscar acknowledged what few other handwriting experts would admit—letter comparisons were not always reliable. As much as he wanted to deliver a solid case against Hightower, Oscar was determined to not help convict an innocent man. For weeks, Oscar examined all the evidence and subjected the clues to a litany of tests using microscopes, cameras, and chemicals. He finally requested a meeting with the police and the district attorney in San Francisco. The prosecutor took notes as Oscar pulled out some large photographs and spread them on a table, each featuring huge rocks. The prosecutor and police chief looked at each other.

"They are grains of sand," explained Oscar. "You see, when I put the hilt of Hightower's knife under the microscope, I found a very small patch of sand. Only a few grains, three or four. They were all I needed."

He had also identified sand embedded by high coastal winds in Hightower's tent found inside his room. But small amounts of sand, even under a microscope, looked similar. Anxious to offer the police conclusive evidence, Oscar did something remarkable. He used a test that he had learned as a chemical and sanitation engineer in Tacoma—a groundbreaking technique in forensic geology that has since changed how police investigate crimes.

Oscar subjected the grains from the tent and the hilt of the knife to petrographic analysis, a microscopic procedure used to help determine the source of the rock that makes up each grain. A petrographer, like Oscar, would use a specially designed optical microscope to identify minerals and rocks within the grains of sand by using polarized light and special prisms. When Oscar switched on the light, the different elements that composed the sand became visible in contrasting colors. He listed each element and compared them to the elements in the grains of sand in the tent—and they matched. It was more valuable circumstantial evidence.

The petrographic test had never been used before on sand in a criminal case anywhere in the world. It was a brilliant innovation, one that is still heavily relied upon in forensics today. It was the first of many adaptations Oscar developed out of necessity, because the field of forensics was evolving so rapidly.

The first mention of forensic geology (aside from *A Study in Scarlet*, an 1887 Sherlock Holmes novel) came from German forensic scientist Hans Gross in 1893, who had suggested in a textbook that studying the soil from a suspect's shoes might be useful in linking him to a crime scene.

In 1904, another German forensic scientist, Georg Popp, was the first to use geology to solve a murder—the strangulation of a seamstress in Frankfurt. Popp placed material found on a handkerchief at the crime scene under his microscope; he spotted bits of coal mixed with grains of minerals and linked the physical evidence to a worker in a coal-burning gasworks who was already under suspicion. It was a landmark case that encouraged other European experts to develop their own ideology about trace evidence.

In 1910, French scientist Edmond Locard developed one of the fundamental theories of forensic science known as Locard's Exchange Principle. "Whenever two objects come into contact, there is always a transfer of material," he concluded. A killer would unknowingly carry trace evidence from the crime scene, and he would likely leave evidence there, too. A forensic scientist was tasked with uncovering those clues.

In 1921, Oscar was determined to use geology to shore up the case against William Hightower. The results of his petrographic test found that the grains of sand on the tent in Hightower's room and on the jackknife in his pocket seemingly matched with the sand at the gravesite. Modern forensic geoscience experts would later laud Oscar, praising him as the first to "extend geological, petrographic investigative techniques to sand, soil, paints and pigments." It was an extraordinary discovery, an inventive technique that made forensic history. He had also proven that some cord found at the gravesite was made of similar fibers as cord found in Hightower's room. But, detectives wanted to know, would Oscar's evidence be enough to send a killer to the gallows?

By early September, the prosecutor had buttoned up his case against William Hightower, but there were still some troublesome details. What if there *was* a second man—a small, dark foreigner? What if Dolly Mason was real?

Reporters following William Hightower's case fixated on his bizarre demeanor, particularly his preposterous story about a missing woman who had snitched on a foreigner with a violent hatred toward Catholics. But soon Hightower's batty behavior would also become a focus of his murder trial. Hightower had hoped to assist the police, he insisted repeatedly, by leading them to the priest's body. They might have never recovered Father Heslin without his help. Hightower didn't seem guilt-ridden or remorseful. He seemed desperate to convince detectives that he would never hurt a priest. Everyone who interviewed him agreed that William Hightower *really* thought he was innocent, but why wouldn't he simply admit his guilt with so much evidence against him? A look at his troubled history offered a few clues.

One year earlier, Hightower wrote to a lender who fronted money for a business that would produce candied fruit. He lamented his recent divorce, his money troubles, and his mental health for the failed venture.

"Five years on the brink of bankruptcy, and losing my wife, left me in a condition both mentally and physically where a rest and a change of scene was necessary," he wrote.

Hightower had certainly suffered from an untreated mental illness, which was likely schizophrenia, but his attorneys never pursued a formal insanity plea. A judge could have assigned Hightower to an insane asylum, a sentence that might have seemed preferable to the gallows. Or perhaps not.

In the 1920s, there was very limited knowledge of psychosis and almost no useful treatments for schizophrenic patients. Mortality rates in asylums were five times that of the general population because of overcrowding. By the 1920s, unqualified doctors were given carte blanche to use chemicals, surgeries, and incarceration to cure mental illness. Psychiatry had not yet entered universities. Hightower surely suffered from a mental illness, but instead of receiving treatment, he faced a murder trial and perhaps the gallows.

And despite his professional successes, in the fall of 1921, Oscar Heinrich was also struggling with his own mental health as he grappled with immense anxiety. Of his knowledge and intellectual gifts, he had little doubt. But his finances continued to be a problem. And despite financial struggles, he had recently signed a lease on a new office in San Francisco at 25 California Street, a much-needed expansion for his thriving business. He was struggling with multiple cases, and his body was suffering from nervous indigestion from the stress.

"I find that I have to make an engagement in court with about three days' work and one day to do it in," he told Kaiser, "that I get along better if I don't eat anything at all as the stress of the work absolutely paralyzes any digestive process."

Oscar's habitual fretting had turned to mania—he would regularly spend twenty-four hours with no sleep as he toiled away in his lab downstairs. Oscar was evaluating evidence and writing reports at a maddening pace. And he was irked by phone calls from rival experts who managed to rile him even more than his noisy boys, who played loudly above his laboratory. For example, the criminalist was infuriated when elderly handwriting analyst Carl Eisenchimmel called an unexpected press conference to brag about his "leadership" on Hightower's case, despite contributing almost nothing of note to the prosecution's case.

"He has hustled around to the newspapers and told them that he was in sole charge of the case on the handwriting end," complained Oscar to Kaiser. "It merely indicates the old man's fear that perhaps out of it I may get some mention which might be equal to his own."

Oscar deeply resented being treated as a subordinate, especially to other so-called experts. "Were it not for the fact that this is a criminal matter of extraordinary interest I would have declined the assignment," he wrote to Kaiser. "I am keeping out of his way as much as possible, hoping that this case will be his final appearance in responsible matters."

William Hightower looked alarmed as the rubber tubes tightened against his chest, and in fact, everyone in the room was concerned. He squirmed as his pulse quickened. Two months before Hightower's trial began, San Mateo's district attorney made an unusual decision, one that would change the way suspects would be interrogated in America. The method was unorthodox, and the machine had never been used before in this context, but the investigators wanted clear, concise answers. They needed them.

This would be a trying day for William Hightower, even before he was strapped to that distressing machine. As prosecutors were compiling their final arguments, one of Hightower's fictional characters suddenly appeared to the shock of police and prosecutors, who were convinced that their suspect was a loon. In early October, a gorgeous, brown-eyed brunette strolled into the Hall of Justice on the arm of her new husband.

"Doris Shirley," she replied when asked her name.

She was indeed real and not a figment of Hightower's wild imagination, but that was the end of the good news for the murder suspect. Shirley denied having any involvement in the killing of a priest. She was not with Hightower on a long drive, as he said, during the night of the kidnapping. She had been preparing for bed at their shared room around eleven when he finally arrived home. She had ruined his alibi, a horrible blow to his defense. And yet William Hightower was still enchanted by the pretty Doris Shirley.

"Whatever she says is all right," he said when told she would not support his story. "Even if she's going to marry another fellow I still love her. Maybe her memory is just short."

Police were assembling a profile of William Hightower, and it was dumbfounding. Was he guilty and brilliant, innocent and unlucky, or perhaps guilty and insane? Oscar believed that there was only one method that could test Hightower's ability to tell the truth: "the apparatus." And the newspapers were willing to pay the bill to discover the results. He called his good friend August Vollmer, Berkeley's police chief. Oscar needed a new tool for this case, and Vollmer might just have the solution.

Hightower had hoped that this cruel day was finally over when a guard awoke him around midnight in his second-floor cell and led him downstairs to a large room. Vollmer and the district attorney stood against the wall as a tall, handsome man in a stylish suit stretched out his hand. He was Dr. John Larson, a fledgling police officer with the Berkeley Police Department, but one with quite a reputation—the first

American police officer to earn a doctorate degree, one in physiology, the study of how a living organism functions. For his master's thesis Larson theorized how fingerprints could predict evil tendencies.

The twenty-eight-year-old was a thoughtful investigator, a learned cop with an educational background in medicine and a fascination with complicated forensics. In the next hour, John Larson would earn another important distinction—the first person to use the polygraph, his own invention, in a criminal case. And William Hightower, in the midst of a swift mental breakdown, would be his first experiment.

Larson's test was divided into three sections: The first part used control questions—name, age, and hometown—to establish normal blood pressure and respiration. The second set of questions was used to monitor reactions to commonplace questions, perhaps a favorite meal or the name of a pet. The third section contained the combative questions about the priest's murder.

As Larson's boss at Berkeley Police Department, August Vollmer had an excellent opportunity to make history. And the results would certainly help Oscar bolster the case against Hightower. The criminalist applauded his friend for moving the needle on improving criminal investigations with this new invention. Oscar's fragile ego was threatened by numerous people, but never by Vollmer. This device would change the justice system, Oscar predicted, and he was pleased to be involved in its first case.

In 1915, Harvard University psychology professor and lawyer William Marston invented the systolic blood pressure test, the first of its kind. It measured blood pressure intermittently, which wasn't especially useful during police interrogations.

Even though Marston was intrigued by science, he was mesmerized with Hollywood; he would later create the character of Wonder Woman, a superhero who brandished the magical golden Lasso of Truth to force criminals to tell the truth. Marston had dreamed that his device could be his own secret weapon, so in 1921, Dr. Larson took Marston's blood pressure test and enhanced it.

Larson asked UC Berkeley's physiology department to develop a new machine, one that would test blood pressure continuously (rather than intermittently), as well as measure heart rate and respiration. Larson and others theorized that those readings would indicate if someone was lying. He named it the "cardio-pneumo-psychograph," but August Vollmer would later dub it a "lie detector" in the newspapers.

By late summer of 1921, Larson had developed a portable polygraph machine using a breadboard as the base; in August, he and Vollmer hauled it forty miles south of Berkeley to Redwood City to use for the first time on a murder suspect.

It had taken Larson a half hour to set up all the equipment before Hightower slid into the wooden chair and put his arm on the table. The suspected killer felt weak—he hadn't eaten in days, and he slept poorly in jail. He was interrogated almost daily.

The media declared Hightower guilty even before trial with headlines like: "Story of Hightower Is Gradually Being Broken by Police." In Redwood City mobs threatened to kill Hightower as vengeance for the murder of Father Heslin, while the police tried to keep him alive for his upcoming trial.

"He will be spirited out of the city, when we feel it is safe," said the district attorney. "We will take no chances with a lynching."

The cornerstone of the defense's case was the memory of a pair of eyewitnesses. Father Heslin's neighbor and housekeeper had both reported that the kidnapper was a short, dark foreigner, not a tall, lanky Texan. But now the women revised their statements—they were mistaken. Father Heslin's housekeeper, Marie Wendel, even offered a melodramatic response when she confronted Hightower in prison.

"My God! It's him!" cried Wendel. "It's the man who took Father Heslin away. The face, the features. Oh, I—!"

The housekeeper collapsed as Hightower stared down at her, slightly confused. Wendel was clearly an unreliable eyewitness, which might not be surprising to modern-day defense attorneys.

Eyewitness misidentification is the leading contributing factor to wrongful convictions, according to the Innocence Project, a nonprofit legal organization that works to exonerate wrongly convicted people. Investigations can be derailed by unreliable identifications, either witnesses who knowingly accuse an innocent person or those who might be traumatized by the crime itself, like the survivor of a deadly shooting. Despite proof that traditional lineups can offer inaccurate results, eyewitness identification is still among the most commonly used evidence in criminal cases. In William Hightower's case, valuable time was lost because of the ethnic bias of unreliable witnesses.

Hightower, haggard and emotional, continued to ramble to the police and the press. He lamented about whippings he had received as a child and sulked over his struggles with writing musical poetry. The press speculated he might be angling for an insanity plea, but that never came.

"My head seems to swell when I think," Hightower told a guard. "It seems there's going to be an explosion. I wonder if I'm going crazy."

The next day he made a bold statement to the district attorney.

"I'm through," Hightower said sadly. "I don't care what happens. I've been unlucky all my life."

It was late that same night, and he was trapped in an uncomfortable wooden chair at the end of a taxing day. Hightower eyed the odd contraption and the two young men standing beside it. The machine had wires, a glass bulb, and two needles suspended just above a wide strip of black smoked paper. Tubes encircled Hightower's chest, and a cuff was strapped to his left biceps. The hands of Larson's assistant, Philips Edson, were smudged from the black paper as he flipped on several switches. Larson knelt by Hightower, his hand pressing a stethoscope against his suspect's right arm.

The needles scratched funny markings onto the paper as the sheet rolled over two wooden cylinders. The three instruments spit out results during the hour-long interview. Hightower answered each query calmly, and then John Larson asked the most important question:

"Did you murder Father Heslin?"

"No," was Hightower's confident answer, but Larson's machine seemed to violently disagree. The scratches became erratic.

"The suspect was covering up important facts on every crucial question asked," declared Dr. Larson to the press. "There were marked rises in Hightower's blood pressure, accelerations and marked irregularities as well, following the vital questions."

August Vollmer called the test "infallible."

"Mere embarrassment or fear are registered in different ways," explained a newspaper report. "Berkeley police say there is but little chance of making a mistake."

That wasn't true, and even in the 1920s, astute judges suspected that Larson's invention might be rubbish. Just two years later in 1923, the United States Supreme Court ruled on the landmark case of *Frye v. United States,* which concerned a murder trial where the suspect tried to have a polygraph admitted into court to help his defense. The court ruled against the suspect, contending that the polygraph had not gained "'general acceptance' in the relevant scientific community." It had not been peer reviewed by researchers or appropriately tested to use in court, and it wasn't known whether it had a high rate of error.

The Supreme Court's decision was superseded in 1993, when the justices decided that all forensic science, including the polygraph, could be admitted in federal courts if it passed new criteria called the Daubert standard. The forensic technique in question had to prove that the "underlying reasoning or methodology is scientifically valid and properly can be applied to the facts at issue." The expert analyzing the evidence must also have valid credentials for using the technique.

Modern scientists agree that there are far too many variables that might control heart rate and respiration, things like mental illness or certain medications. And there is no marked difference between embarrassment, fear, or anxiety. August Vollmer was wrong, and yet, since he introduced the invention in 1921, polygraphs have been used in countless criminal investigations, including federal cases. There are people sitting in prison right now because of one piece of junk science that was pioneered by a good cop.

Guilty, concluded the men controlling the machine attached to William Hightower in 1921, and when he returned to his cell, he collapsed in near hysteria. Oscar wasn't surprised—he knew August Vollmer (and science) could help him shore up his case against a priest's killer.

"I have been thinking too much lately," William Hightower sadly told his attorney. "I get the queerest notions now. I feel sort of lonely."

In his jail cell before his October trial Hightower reflected on his life—his lost loves, his business failures, and his murder charge. His one stroke of luck was the addition of a new attorney, E. J. Emmons, who volunteered to represent Hightower pro bono. He had grown fond of the kooky baker when Emmons watched Hightower give away cookies to the children in Bakersfield, where he said he once owned a shop.

"I don't want to give up the dearest of all things, freedom," Hightower lamented. "But they've got me bound up by a chain of circumstantial evidence so strong I will never be able to break it."

He seemed to be right, because when his trial began on October 5, the district attorney presented a list of clues that all pointed to Hightower as the killer. The hearing began with the grim story of the midnight search for the priest's body as told by newspaper reporter George Lynn on the stand. The trial lasted less than a week, and during that time, the prosecutor paraded more than a dozen witnesses through the court, each adding to the chain of circumstantial evidence against William Hightower. The housekeeper and the neighbor, the only two witnesses on the night of the kidnapping, were scheduled to take the stand to positively identify him, but the neighbor had since left town, leaving Wendel as the only witness.

"There was good light," said housekeeper Marie Wendel. "The

man wore dark glasses. He refused to come into the house, but I could see him plainly."

There was now no mention of the small, dark foreigner. The prosecutor dismissed that misidentification as stress, not xenophobia.

One man said Hightower had rented a 1920 model Ford touring car with a self-starter on the night of the kidnapping and he had been gone for several hours. Witnesses placed Hightower near Salada Beach less than a month before the murder, and one week afterward he had been loitering near the flapjack sign. A man in Nevada testified that he sold Hightower a .45-caliber revolver. It had been a bewildering day for William Hightower, but what would happen next was just simply sad.

"I call Mrs. Lee Putnam as the next witness," announced District Attorney Franklin Swart.

The courtroom stirred. The newspapers had made it clear that this woman, Doris Shirley (now Doris Putnam, as she had recently married), was Hightower's former sweetheart. The twenty-four-year-old sat demurely in the witness chair, perfectly hatted and clutching a small purse. She was dressed in a black cloak, a brown dress, and dark silk stockings. Shirley faced the jury with the glamour of a Hollywood actress as her new gold wedding band glowed. Hightower still thought she was beguiling—he seemed skittish as she sat there. He ripped paper into small bits or scribbled quickly on a pad before suggesting that his attorney ask his former lover a specific question.

"No," Emmons firmly replied, "it would be unwise to ask that."

Shirley described how she met Hightower in Utah two months before Father Heslin's murder and then they traveled to San Francisco, where they lived together for about a month. The day before the priest's murder, Shirley had rendezvoused with Lee Putnam, the man she would marry just a month later. On the day of the kidnapping on August 2, Shirley was supposed to meet with Hightower, but instead she went to the theater with Putnam. Hightower's devotion to her had been futile—she was in love with someone else.

Shirley quickly renounced her former lover, denying his claim that they drove to San Jose and back on the night of the kidnapping. Reporters noted how dejected he seemed as he furrowed his brow and shook his head at her.

"I left San Francisco with Putnam because I wanted to get away from Hightower," Shirley testified.

When his defense attorney began a coarse cross-examination, Hightower tugged at Emmons's arm and demanded restraint. As Shirley ambled past his table toward her new husband, Hightower whispered softly, "Your memory is woefully short, little girl."

On October 11, Hightower took the stand in his own defense, and as expected, he was a terrible witness. He rambled and then repeated those ramblings until he was interrupted. Hightower was clearly outmatched, and his testimony severely damaged his defense. The forensic evidence was also presented and, while a revolver was important, a knife would be the key clue.

"Bad Day for Hightower," declared one local newspaper headline. "Little Grains of Sand May Be Big Factor in the Conviction of Man at Redwood City."

On the third day of the trial, Oscar Heinrich stepped on the stand and presented his case against William Hightower. He guided the jury confidently through each step of scientific reasoning and procedure. He showed them huge photographs of grains of sand, describing the differences between them: color, texture, size, and grain. He explained how a petrographic test revealed the variations in the mineral composition and organic materials in the sand, even its luster. Those small differences, Oscar said, told a story about the two samples of sand in William Hightower's case. The grains found in Hightower's room and the sand from Father Heslin's burial site on Salada Beach most likely came from the same place, an important piece of circumstantial evidence.

During cross-examination, Hightower's attorney questioned the science—he pointedly asked Oscar if all the sand on the Pacific Coast was similar. Oscar replied that he didn't know, but he *did* know that there was a variation in types of sand, and he could say that both samples of sand appeared to come from identical sources—a detail that created a stronger case against Hightower.

"This evidence was considered the strongest produced by the prosecution to connect Hightower with the murder of Father Heslin," wrote one reporter.

The prosecutor walked over to Oscar and asked about handwriting analysis, his assertion that Hightower was a baker based on the unique writing on the ransom letter. Oscar pointed to two huge photographs, each with the letter *D*. One came from the ransom note and the other from Hightower's own writings.

"Look at that letter D," he told the jury. "The particular feature connecting the two writings, a feature that at first glance did not appear, lies in the inability of the writer to make a perfectly straight downward stroke."

Jurors took notes and stared at the massive pictures as Oscar pointed to another set of photos.

"Now the S shows a very distinctive tallying on three points," he said. "The feature of the capital S as found in the ransom letter is the initial and terminal stroke. There is a very highly individualized movement consisting of a horizontal loop made from right to left."

The jury listened closely and stared at the loops—Oscar was mesmerizing as he looked each juror in the eye. The criminalist's forensic work was determining William Hightower's fate.

After spending two days on the witness stand, Oscar was no longer needed in court, which was welcomed news because he was absolutely spent. He had not written a letter to John Boynton Kaiser for weeks, an unusual deviation in their schedule. They were devoted pen pals, codependent confidants who filled a void with each other that neither seemed to quite understand. Their marriage to their wives, while ostensibly loving and stable, appeared to be banal at times. But these two pedantic, uptight men always seemed to loosen up around each other, both in person and on paper. When Oscar failed to send one of his witty notes for almost a month, Kaiser grew concerned. The librarian mailed a casual letter to Oscar meant to provoke a quick response and maybe even a little panic from the reliant criminalist.

"Farewell to criminals and their detection, for the time being," joked Kaiser about his challenging new position at the library. "Fortunately I have sent you enough references to keep you busy till you retire a millionaire."

Kaiser wasn't exaggerating. By 1921, Oscar had received hundreds of books from him, and virtually every one had been used in his criminal cases. Kaiser had spent years handpicking books just for Oscar—the criminalist suspected that the librarian was a bit jealous of Oscar's seemingly glamorous life as a scientific detective.

"Our copy of Lucas 'Forensic Chemistry' came yesterday and I spent several hours looking it over last night," wrote Kaiser. "On my scientific mind it makes a very good impression from the standpoint of scientific completeness and accuracy."

But there was still no response from Oscar, who was in the midst of the Hightower trial. And that really seemed to irk Kaiser. He was hoping to collaborate with Oscar on a series of articles about using science labs to solve crimes, so he sent his friend a short list of potential titles—all dry as dust, unfortunately. Crafting a unique, catchy title for an article aimed at the layman was almost an art form, and Kaiser wasn't quite gifted at it.

Inside his lab, Oscar thumbed through the letter and paused at one sentence. Now *he* was irked. He stared down at a name: "Chauncey McGovern." Kaiser had made an innocent gaffe when he encouraged Oscar to read a fascinating *Literary Digest* article entitled "A New Way to Trap Forgers," written by Oscar's acerbic foe.

"It occurs to me that you may wish to write the *Literary Digest* sometime on this subject yourself pro or con and commenting on Chauncey McGovern's work," the librarian wrote.

Kaiser had likely forgotten how much Oscar despised McGovern. The criminalist ignored the letter. When more than a month went by with no response from his friend, a frustrated Kaiser shot off a telegram, and Oscar finally replied, but not kindly.

"Your idea on articles which you have suggested is not bad," Oscar wrote. "Your titles however are rotten, absolutely too ponderous to get anything over . . . not as the title for a short story. You'll do better to head it 'How to Shoot a Husband and Get Away With It' or something like that."

Oscar's complicated murder cases, his financial concerns, along with his competition with McGovern, were all straining their friendship. The criminalist didn't seem to appreciate being pressured by his best friend. And he certainly disliked being reminded by Kaiser of all people that McGovern was considered a legitimate expert. Kaiser was wounded but tactful with his response.

"Your unflattering comments on the titles I suggested carefully noted," Kaiser delicately replied. "However you're all wrong. I was contemplating serious articles and not merely short stories."

Kaiser's response was mature until its finale—he couldn't allow Oscar to win a battle of egos *every* time. Weeks earlier, the librarian had convinced Oscar to buy a Dictaphone for recording dictations because the forensic scientist had often complained about his secretary's poor typing skills. Kaiser mocked him in one final jab that was sure to injure the criminalist, even if it was draped in sarcasm.

"By the way, I wonder whether you have a real Dictaphone voice," Kaiser said. "There are so many errors in your letter."

Bickering was not unusual for the pair, but insults were rare. However, repairing Oscar's friendship with Kaiser would have to wait, because Hightower's defense team was presenting closing arguments.

William Hightower's attorneys were hobbled from the beginning. The judge refused to grant them most of their requests, and he frequently sided with the prosecutor during objections.

The jury retired around two o'clock on October 13, and as Hightower watched the jurors exit, he chewed bubblegum, a habit noted by reporters from the moment the trial began. It seemed to calm him, but it was exasperating to hear. His attorneys hoped that, if there had to be a guilty verdict, Hightower would be spared the death penalty. It was a gamble, because a man who killed a priest certainly didn't deserve mercy, according to most Americans, but Hightower was so peculiar—several notches above eccentric.

About an hour after the chamber door shut, the jurors requested the ransom letter. About fifteen minutes later, they sent for photographs of Hightower's own handwriting. They spread them all out on the table, debated, and compared. And then they discussed Oscar Heinrich's testimony about the sand, the tent, the fibers, and the ransom letters. Each clue, combined with the others, had proven that Hightower was at the murder scene. The jury returned with a verdict in less than two hours. As Hightower stood and faced the jury silently, he continued to chew gum.

"We find the defendant guilty of first degree murder," said the foreman.

Spectators whispered at first, and then the sounds grew louder before the bailiff shushed the room. Hightower showed no emotion but continued to chew. He listened and then turned to his attorneys and some reporters nearby.

"Well, boys, I guess you won't see me for some time," he said dismally.

But there was one surprise, a remarkable decision from the jury. The panel recommended life imprisonment for William Hightower and not the gallows, even though he had murdered a priest. He would spend the rest of his life in San Quentin State Prison. Oscar Heinrich was often a proponent of capital punishment in particularly heinous crimes, but he agreed that mercy in this case was appropriate.

"In my opinion the case got away from Hightower," Oscar wrote to Kaiser, after the two had reconciled. "When the priest stuck his head in the tent and found there was nothing there, he put up a fight."

Oscar reminded Kaiser that Father Heslin was a tall, burly man, one certainly capable of overpowering a lanky kidnapper like William Hightower.

"The priest was so powerful that I am satisfied that he was killed in this fight with no original intention on the part of Hightower of doing him any serious injury," concluded Oscar.

And he was proud that he had been able to contribute to the case.

"The knife connected him definitively with the tent and the grave and was the big thing in the case," he explained. "According to the jury it was the particular thing which brought about the conviction."

But there were still concerning questions in William Hightower's case—he had always maintained his innocence, despite the evidence. He had never wavered.

"Regardless of what the jury and the public may think," Hightower said, "I am innocent. Yet I knew I was going to be convicted."

One of the most troubling questions involved motive—if William Hightower had kidnapped the priest for ransom money, why did he kill him before negotiations even began? If it had been a hate crime against a Catholic clergy, why send a ransom note at all, and why lead the police directly to the body? There had been nothing to connect him to the case until he approached the archbishop's mansion. Oscar speculated that it was a type of atonement for Hightower—he had hoped to be caught. Or maybe there was no real answer, just a tragic diagnosis.

William Hightower turned out to be little help with those questions. He lived as a recluse inside San Quentin, working in the prison's furniture factory and writing poetry for almost forty-four years—four decades without treatment for a mental illness. He lost each of his appeals, but in 1965 he was released at age eighty-six, the oldest inmate of the California prison system.

"I have no feelings, no bitterness against anybody," he insisted. "I'm going out."

William Hightower died just a few months later in a halfway house, alone and unrepentant. He never admitted to killing Father Heslin.

Oscar Heinrich's stomach churned again, a visceral reminder of his stress. All of America now knew that he was the star witness in William Hightower's murder trial—E. O. Heinrich, as readers knew him, had doomed the monster who murdered a beloved priest. Oscar was confident, buoyed by well-deserved public accolades. He sifted through dozens of articles, each one mentioning his name. He adjusted his spectacles. His head ached occasionally when he missed an afternoon nap. But there was another reason.

Oscar told few people, maybe even just Kaiser, that over the last month he had felt nearly crushed under the strain of *two* remarkable investigations, not just one. With William Hightower finally convicted, Oscar Heinrich could now concentrate on the most notable case of his career—one that had begun a month earlier, while Hightower was still awaiting trial. The criminalist looked over at a thick manila folder and eyed its handwritten label. It read: "Fatty Arbuckle."

5.

Damnation:
The Case of the Star's Fingerprints, Part I

As he held the waxen print close to the blood-stain, it did not take a magnifying glass to see that the two were undoubtedly from the same thumb. It was evident to me that our unfortunate client was lost. . . . "It is final," said Holmes.

—**Arthur Conan Doyle,**
The Adventure of the Norwood Builder, **1903**

Virginia Rappe was dying. Her abdomen ached on Thursday, September 8, 1921. The twenty-six-year-old was lying in a posh hotel room, one that managed to make her feel both spoiled and distressed. A showgirl named Maude Delmont drifted around her. By then, the older woman no longer reeked of alcohol, but it had taken more than a day for the rancid smell of whiskey to disappear.

Three days earlier, Rappe had collapsed at a party in San Francisco, and now Delmont was comforting her. They were acquaintances, not confidants, but Delmont was the only one who volunteered to stay. Rappe moaned, held her stomach, and whispered secrets to the showgirl.

At least three doctors hovered over her hotel bed, pressing her abdomen and prodding her body. They questioned Rappe, and she had tried her best to respond despite the morphine. The physicians gently examined her, but they found no signs of injury, no telltale evidence of sexual assault or physical abuse except for a few minor bruises. They asked about the party—Rappe admitted she had consumed orange blossoms, a drink of equal parts gin and orange juice, which was sometimes blended with sweet vermouth or grenadine. There were bottles of scotch, gin, wine, and bourbon flung around an adjoining suite—all of it illegal.

The doctors offered her no treatment other than a warm compress and some opium for the pain. There was no need to take her to the hospital, they assured Maude Delmont, because there was little to be done for alcohol poisoning. The stylish brunette's body twisted as Delmont relayed the story.

Rappe had been one of the leading ladies at a bash in the suite next door, a drama starring five men with ties to Hollywood and four showgirls who were colored in heavy makeup. They were all there to celebrate one of the most affluent, influential Hollywood actors in the early 1920s—Roscoe "Fatty" Arbuckle.

Fatty (who hated that nickname) was larger than life. In fact, he was larger than most actors, weighing more than 250 pounds. Arbuckle's cherubic face grew full when he laughed, round and bright with a double chin and red cheeks.

Born in 1887 in Smith Center, Kansas, Roscoe Arbuckle struggled right from the start. He was a large baby, and because the rest of his family had slim builds, his father assumed that he was not biologically his child. Roscoe's father came around, but he never let his son forget the doubts about his paternity. After a few years, the Arbuckles moved to Santa Ana, California, where Roscoe developed his singing voice, eventually finding his way to the stage and in films after his mother died when he was twelve. He was hilarious, a brilliant comic and actor who grew more famous with each performance.

As the most popular silent film star in 1921, Fatty Arbuckle commanded respect and reverence, despite his proclivity for playing buffoons on-screen. Audiences lined up to see his comedies so often that

Paramount Pictures signed the thirty-four-year-old to an unprecedented $3 million contract to star in eighteen silent films over three years, an incredible fee for a comedian who often played a naïve hayseed.

Arbuckle was a kingmaker, of sorts. He had discovered Buster Keaton and Bob Hope—he even mentored Charlie Chaplin. Arbuckle had been married for thirteen years, but he and his wife, actress Minta Durfee, had recently separated. Arbuckle enjoyed a hectic shooting schedule, one befitting a power player in Hollywood.

His latest film, *Crazy to Marry,* had been released a week earlier and was playing in theaters across the country, so his friends insisted on throwing him a party in San Francisco, a shindig at an exclusive hotel on Union Square that seemed pulled right from a glamorous film set.

Built in 1904, the St. Francis was *the* hot spot for celebrities and literati of the Jazz Age to socialize, a prime location for an elegant soiree featuring luminaries like Chaplin and Sinclair Lewis. Styled after Europe's most glamorous hotels, with intricate wainscoting and detailed trim, a room at the St. Francis would have been the perfect respite for a peaceful evening—but it was a dreadful place to die.

Virginia Rappe was a former fashion model, a minor actress, and an accomplished clothing designer whose career seemed to be sliding to a stop. She was also a party girl, a regular at Tinseltown affairs who was popular for her blatant flirting and stunning looks. Rappe enjoyed drinking with celebrities, including Arbuckle. The cigarette smoke from the party had faded days earlier, though, and now Rappe was suffering in a dark room, surrounded by a trio of nurses who all blamed liquor.

Rappe was moved that afternoon to a nearby medical facility, examined by more doctors, and given a new diagnosis: peritonitis. Her abdominal lining and cavity were severely inflamed from an infection. Physicians would later discover that she was afflicted with chronic cystitis, a recurring bladder infection made worse by large amounts of alcohol. But that's not what would kill her. Her bladder had ruptured, they determined, because of "some external force." As she lay dying, Virginia Rappe offered her final words.

"Oh, to think I led such a quiet life," she cried. "And to think I would get into such a party."

She died the afternoon of September 9, 1921—a starlet with so much potential who would soon monopolize newspaper headlines for a horrible reason. After her death, the police were called. Detectives began interviewing witnesses, gathering information about exactly what happened on the twelfth floor of the St. Francis Hotel. There were other showgirls at the celebration in addition to Maude Delmont—women with bit parts in life who were now feeling the glare of a very uncomfortable spotlight. Alice Blake recounted her version of the party to both investigators and the district attorney, a murky story about a drunken silent film star who couldn't control his lust. Newspaper reporters, tipped off about a potential star-studded scandal, called Fatty Arbuckle at his Tudor-style mansion in Los Angeles even before detectives arrived. He recounted a candid story about a woman who suddenly snapped.

"Miss Rappe had one or two drinks," Arbuckle explained. "She went into the other room of the apartment and began tearing her clothes from her body, and screaming."

She complained of breathing problems, so two women in the suite lowered Rappe into a tub of cold water while Arbuckle secured another room. After she calmed down, Arbuckle and another party guest, actor and director Lowell Sherman, moved her there and phoned a hotel doctor.

"After he reported that she had quieted down, Sherman and I went down into the dining room and danced the rest of the evening," said Arbuckle.

The actor had no idea that her illness was serious; otherwise he would have stayed, he assured detectives. The investigators glanced at each other and began asking more pointed questions about their relationship—how long they might have been alone in the room together.

Arbuckle grew quiet and then concerned. Modern filmmaking in Hollywood, according to some Americans in 1921, was the devil's work.

Movie scenes were becoming uncomfortably risqué; women wore less clothing while men used curse words and sexually suggestive language. In 1919, director Cecil B. DeMille presented *Male and Female,* a controversial film that explored gender relations and social class—and infuriated conservatives. The morals of the country seemed irrevocably bound to Hollywood, and that was dire news, according to many.

The details of Fatty Arbuckle's party, rumored to be fueled by sex and gin, seduced the American press, to the delight of religious leaders. William Hearst's newspapers declared that Hollywood was a modern-day Gomorrah, temporarily transplanted to San Francisco for one evening. Reporters fixated on Virginia Rappe's character, harping on her multiple broken marriage engagements. They dissected her outfit that night, her hairstyle, and even her voice. And Fatty Arbuckle's supposed voracious sexual appetite was detailed in print. Soon many moviegoers would begin to mistrust the country's favorite on-screen comedian, a box office draw like no other actor—a legend at age thirty-four. The comedian was transposed from a lovable media darling to a murderous sexual predator in less than a week. Rigidly moralistic community leaders who had fiercely protested the vices of Hollywood now demanded the noose once they read accounts from women at the party.

"I am dying! I am dying!" witnesses said Rappe cried that night.

"We heard Miss Rappe moaning," said showgirl Alice Blake, "and Arbuckle came out of the room."

That night Fatty Arbuckle was ordered to San Francisco for a talk with police.

Showgirl Zey Prevon surveyed the gentlemen in the room. She always felt skittish near police, so she was frazzled for much of the day as she awaited this interview about her role in Fatty Arbuckle's now notorious party.

Prevon's real name was Sadie Reiss, but she used a variety of stage names to garner attention from film producers. This was not the audience she had hoped to attract that night. It was a worrisome mess for a volatile woman who was prone to histrionics and craved the spotlight. Soon she would receive loads of attention—none of it good.

Prevon explained that she had arrived to the suite in the St. Francis Hotel on Monday, September 5, at about one thirty p.m. after leaving her roommate in the lobby. Arbuckle and Lowell Sherman, the director, were both dressed in robes and pajamas. And soon Arbuckle and Virginia Rappe disappeared into his private room, suite 1219.

"How long did they remain in there?" asked Assistant District Attorney Milton U'Ren.

"A good long while," replied Prevon. "I went over and banged three or four times on the door."

Prevon and Maude Delmont pressed their ears against the thick wooden double doors, straining to hear anything, but there were no screams, no raised voices. The showgirls frantically knocked, demanding that Arbuckle let them inside to check on Rappe. The pair had been sequestered in his bedroom for around an hour, Prevon guessed. Arbuckle slowly opened the door and reluctantly allowed the women inside, fixing his bathrobe as they pushed past. They stared down at Rappe as she clutched her stomach.

"She was lying on the bed. Her hair was all down and she was moaning," said Prevon. "'Oh, I am dying.'"

The actress was fully dressed, according to Prevon. As Fatty Arbuckle watched Rappe writhe on the bed, he glared at the showgirl and snapped.

"Get her out of here," Maude said he yelled. "'She's making too much noise.'"

Prevon was stunned and then concerned. She told police that when Rappe began screaming and pulling off her own clothing, Arbuckle tried to remove them for her.

"I said, 'Don't do that, Roscoe,'" testified Prevon. "I said, 'She is sick.' He said, 'Oh, she's just putting on.'"

Prevon said that Alice Blake and Maude Delmont stood by the bed and tried to calm Rappe by offering her warm water and bicarbonate of soda to settle her stomach; soon they slid her into a cold bath, like Arbuckle said. Prevon tried to ring the hotel doctor, but someone in the suite snatched the phone.

"They couldn't afford the notoriety," she explained.

Arbuckle seemed agitated as Rappe's screams grew louder; his attitude vexed Prevon, so she confronted him.

"Oh, if she makes one more yell I will take her and throw her out of the window," barked Arbuckle.

"Did she accuse him of anything?" asked Milton U'Ren.

"She was just yelling, 'I am dying; I am dying,'" recalled Prevon. "'You hurt me.'"

The gentlemen in the police interview room grew silent. That was a severe accusation, one that could result in a murder charge. Prevon leaned over the typed document on the desk and added her signature. Detectives warned her against talking to anyone—she would be called to testify against Arbuckle.

"We don't want people running to you and all that kind of thing to have you change your story," cautioned Captain Duncan Matheson. "They will."

"Well, I won't," she replied.

Fatty Arbuckle arrived at San Francisco's Hall of Justice after the women left the police station on Saturday night, September 10. He sat in a wooden chair and listened closely. The assistant district attorney tried to question him, but on the advice of his attorney, the actor refused to respond. Arbuckle's silence infuriated the captain of detectives, and he vowed to uncover what really happened in suite 1219.

"Neither I nor Mr. U'Ren nor Chief of Police O'Brien feel that any man, whether he be Fatty Arbuckle or anyone else, can come into this city and commit that kind of an offense," said Duncan Matheson. "The evidence showed that there was an attack made on the girl."

The district attorney accused the actor of sexually assaulting Virginia Rappe and then accidentally crushing her under his 266-pound frame. Police arrested him, charging him with murder.

Newspaper reporters were thrilled with a new tabloid story, a horrible death starring an iconic entertainer. Arbuckle's arrest would be the opening salvo to a media circus never before seen in Hollywood. And soon Oscar Heinrich would be at the center of the movie industry's first major scandal, a case that would forever reshape his reputation—and not for the better.

David "Kid" Bender eyed his new neighbor as the man paced his six-foot-square room, Cell No. 12 on "Felon's Row" in the city jail in San Francisco. The new inmate had been there just seven hours, but Bender could tell that he was already miserable. At seven the steel doors swung open, and Bender wandered into the narrow hallway that separated the two rows of cells made from solid steel walls. There were eighty other prisoners, and almost all of them were whispering about the new tenant in the cell adjoining Bender's room—Fatty Arbuckle. The cop killer strolled over to the cell's opening and stood nearby as Arbuckle walked out.

"Has anyone any soap?" Arbuckle called out. "And a towel and a comb? I haven't anything. Not a thing. Nothing."

Bender watched as the actor turned out his empty pockets. The twenty-three-year-old offered to loan Arbuckle the supplies as they shook hands and smiled. It was ironic, of course. Just one week earlier, the actor had been chatting with wealthy movie executives, plotting his next career move. This morning he was surrounded by violent felons and chatting with an escaped murderer from Maryland who had never even watched one of his films.

"They don't show many pictures where I have been for the past six years," Bender joked to Arbuckle.

But Bender and Arbuckle did manage to find some common ground: they were both charming, both intelligent, and both misunderstood (they believed). And there was one other thing.

"We ought to be friends," Bender said with a smile. "One of the girls in your party has been living at the apartment where I stayed."

Arbuckle's face fell. He groused quietly. Unfortunately, the actor did know David Bender's roommate—it was Zey Prevon, the woman who had knocked on the door as Virginia Rappe lay dying. Arbuckle felt helpless. A small chorus of showgirls was each lining up to testify against him—three women who had convinced the prosecutor to charge the actor with murder. It was such a peculiar, intimate world when criminals and celebrities mingled in San Francisco in the 1920s.

"I'm through with the booze," Arbuckle bellowed on Felon's Row.

On September 16, Oscar Heinrich listened as his secretary greeted a caller, San Francisco's assistant district attorney Milton U'Ren. Despite being inundated with evidence from the Father Heslin case, which was running concurrently, the forensic scientist was excited about this new investigation. Oscar would be the state's lead criminalist in the case against Fatty Arbuckle—no more playing second fiddle to arrogant handwriting experts. The next morning Oscar reported to the Hall of Justice to meet with District Attorney Matthew Brady; it quickly became clear that this case would be knotty. Brady advised him to keep a low profile.

"I went to work this morning—incog[nito]—on the Arbuckle case," Oscar told John Boynton Kaiser, cryptically. "Am living at the St. Francis for a few days. As long as I dare, will give you some details."

On Friday, September 16, Oscar and his assistant Salome Boyle greeted the police officers at the St. Francis Hotel around one in the afternoon. They hauled along loads of equipment: small bags for collecting evidence, tweezers, microscopes, a magnifying glass, and drills for removing doors. Boyle lugged along a high-powered light. It was five days after prosecutors had charged Fatty Arbuckle with murder, and this was Oscar's first day of exploring the now-notorious suite 1219 and its adjoining suite 1220. He would return three more times before declaring the clue-gathering phase complete. Just a few hours earlier, U'Ren gave him specific instructions.

"Make scientific and microscopic examination of marks on door between 1219 & 1220," he wrote in his field journal, "particularly of inside of 1220."

Oscar faithfully, meticulously filled out several pages of his large field journals every day of the week, even on the weekends and holidays. He chronicled specific times for every appointment, phone call, or scientific test and noted the case involved in the margins. He required that his secretaries and assistants do the same and, if they refused, he fired them. Oscar always noted when he awoke in the morning, when he fell asleep, and when he required his afternoon naps (almost daily). He even journaled when he journaled, the mark of a fastidious, rigid man: "8pm–10pm, journalizing," he wrote in one entry.

At the St. Francis, Oscar locked the doors behind him and asked his secretary to plug in the portable light. He surveyed the doors first, squinting at each mark and impression.

"Found marks on door between 1219 and 1220 on 1220 side together with footmark showing door had recently been kicked by woman's foot," he wrote. "Reason—varnish dust from abrasion fresh in color; door had not been wiped since struck."

He loved being at the site of the "action" (as he called it)—in the field and exploring a crime scene. He squatted on the ground and began harvesting the rooms for clues. Oscar used thin metal tweezers and carefully collected two hairpins, a difficult task because the carpet was a tightly woven, dark material with a busy pattern. In the glare of his laboratory light, tiny strands glowed. There were dozens of hairs, some pubic, which required hours to collect. He later gently taped them to a sheet of paper and then used his ruler to record their measurements. He typed out notes about everything: measurements of the room, sizes of the furniture, even the coordinates of the locations of evidence in relation to the opening of the doorway: "1. Ordinate 41 inches, Abscissa 56½ inches, Long reddish or golden hued woman's hair."

There seemed to be dust everywhere, as if the rooms had been neglected for several weeks. Oscar ordered the suite to be sealed, a tardy response to prevent a long-ago compromised crime scene. The criminalist worked slowly, making a grid search of both suites. He placed markers on the floor where evidence had been discovered. The pattern of the clues told him a story, a clear narrative about Virginia Rappe's night with Fatty Arbuckle.

"Found evidence of a struggle between man & woman," he wrote in his field journal.

He stared at his most compelling evidence, two latent handprints a few feet above the doorknob on the door of 1220. They appeared to belong to a man and a woman—the man's print was pressed on top. He sprinkled fine powder against the dark door. The cracks in the wood filled with white dust, and Oscar's hidden clues were suddenly exposed. He quickly jotted down complicated math, the potential number of people who had been in that room since the hotel opened almost two decades earlier: "720 per year, 12 years—8640 people may have used room since opening."

He rolled over a metal stand holding his large camera and snapped photos. There were clearly two hands, but the fingertips seemed longer than they should have been; the whirls, arches, and loops were all there, but they were elongated. They weren't simply placed there but dragged, and now they were smudged. He wrote down to "practice developing latents." He also noted that he should call his good friend August Vollmer for help. This case seemed too daunting to go it alone.

Oscar urgently needed to see Virginia Rappe's hands, so in less than two hours he and his assistant were hovering above the dead woman's casket at the undertaker's parlor, preparing for a morbid task. Oscar unpacked some paper along with a small metal roller and black ink. He gently picked up Rappe's hands, one at a time. He pressed her palms onto a white sheet, an official fingerprint form from the San Francisco Police Department. He repeated the process with each of her fingers—now he had samples to compare with the latent prints on the door. He flipped over a notepad and jotted down the measurements of her fingers. He would collect Fatty Arbuckle's prints in jail later.

Oscar picked up a suitcase filled with Virginia Rappe's clothing from the night of the party: a jade skirt and a jade sleeveless blouse over a white silk shirt, completed with a white Panama hat. Inside Oscar also found a pair of panties and two garters.

When he returned to his lab in Berkeley, he used tweezers to carefully seal each clue in sanitized containers. His approach was quite a deviation from the haphazard methods used by most investigators at the time, who refused to use labels and often stored evidence together without regard for order and method.

Oscar lived for order and method.

Oscar and his assistant retired to Berkeley for the night. They returned to the hotel the next morning to search for more fingerprints, but when Oscar and Boyle arrived at the St. Francis that Sunday, someone else had already been there.

"On entering room in morning, found that room 1219 had been entered after my departure and inspection made of fingerprints developed on hallway door," he wrote. "Markers on floor kicked about, but no other damage."

Someone was watching Oscar Heinrich.

The district attorney had an admittedly weak case against Fatty Arbuckle, an investigation that relied on testimony from problematic witnesses like Alice Blake, Maude Delmont, and Zey Prevon.

"As everyone knows, I had had quite a number of drinks myself," Prevon told the press. "But they didn't blind my eyes to what was happening. Virginia Rappe's condition woke me up. Like 'Fatty,' I am off the booze forever."

Oscar and his forensic investigation would have to serve as the linchpin to Brady's case, and luckily for the district attorney, the criminalist's profile was rising in the newspapers thanks to Father Heslin's case.

San Francisco's police chief sent a large lock of Rappe's hair to Oscar's Berkeley laboratory for analysis. And now there were more disclosures. Federal agents claimed to discover an underground "booze" railroad—a network originating in Hollywood and ending in certain affluent San Francisco hotels, keeping them well supplied with the illegal alcohol that fueled wild parties such as the one at the hotel that night. The country's attorney general, Harry Daugherty, and his team were investigating Volstead violations, and they had already received statements from the Hollywood men at Arbuckle's party concerning the illegal liquor. The federal government threatened the guests with arrest and felony charges.

"A regular system was in operation whereby certain wealthy men in the know could come from the southern 'picture and millionaire' colony [aka Los Angeles] to this city," said Daugherty's assistant, "and be assured of ample liquid substance."

In the 1920s, Hollywood had likely never been described so politely. When reporters asked Oscar if he had found evidence to prove Fatty's guilt, he replied yes.

"Following orders from Brady, Heinrich refused to divulge the nature of his findings," read one report, "but intimated that he had found evidence of positive value in the room."

Oscar was confident in Arbuckle's guilt, and Vollmer could take some of the credit. The police chief stood next to him, staring at the suite's door. Vollmer pointed to one handprint in particular, and holding up a photo of Rappe's hand, both men concluded that it belonged to her. Oscar trusted him implicitly.

Oscar was even more explicit in his updates to his friend John Boynton Kaiser. The country was struggling with a crime wave, which included a disturbing increase in sexual assaults. Oscar claimed he knew the cause—and it wasn't organized crime, whorehouses, or Prohibition.

"I perceive a direct connection between the movies and these crimes," Oscar complained to Kaiser. "There is no form of caress which is not displayed in the movies."

He bemoaned the vulgarity of young people and their degrading morals, particularly those of women. Oscar had joined the ensemble of Fundamentalists who cursed Hollywood.

"The prevailing opinion among the college girls of the University of California today is that young men will not dance with them if they wear a corset," Oscar grumbled. "The movie has become such a powerful factor in liberty and license."

Oscar Heinrich was disgusted with the film industry and distressed by young people who mirrored the vile behavior they saw in movie theaters. He complained to August Vollmer, his preferred consultant (aside from Kaiser). Vollmer agreed—he was disgusted with Hollywood's fondness for portraying police officers as clowns.

The criminalist was suspicious of Fatty Arbuckle—a dangerous stance for a scientist who had pledged to remain unbiased during a criminal investigation.

Oscar Heinrich quickly peeked over his shoulder. Someone was watching him as he traveled around San Francisco on September 20. He had just met with the district attorney at the Hall of Justice for a discussion about courtroom strategy in Fatty Arbuckle's trial, and now a man was tailing Oscar, lurking just a few blocks behind.

"I think I lost him about two o'clock," said Oscar in a letter to librarian John Boynton Kaiser. "He picked me up at the Hall of Justice and stayed with me as long as he dared, but he realized that he was caught and dropped out."

Oscar Heinrich had proven to be the state's most important witness, the expert who would alter the course of Fatty Arbuckle's life in a surprising way. Oscar's investigation would make headlines quite soon.

"Have made a number of important discoveries these hicks around here have overlooked," he promised Kaiser.

Oscar's letters had frequently amused the reference librarian, especially when they contained gossip about cases or criticism of dim cops. The scientist and the librarian both considered themselves to be intellectual giants with little tolerance for dimwits.

"I should think you might get a good deal of fun out of finding yourself shadowed," Kaiser joked.

The forensic scientist enjoyed briefing his closest friend on shocking cases—detailed facts followed by frank commentary. Oscar confided to Kaiser that while he was making progress, Arbuckle's team had stymied him. Kaiser offered an idea. He told Oscar to have a chat with the people hiding in the background at the St. Francis—the housekeepers. Those quiet workers often kept the best information, Kaiser promised.

"You suggest that I chat with some of the maids around here. Those maids are coo-coo," Oscar told Kaiser. "There has been considerable Arbuckle money floating around so I don't think they will talk any more than they have to."

He would regret not taking Kaiser's good advice.

Oscar Heinrich felt nearly crushed under the strain of two remarkable investigations. He was evaluating evidence in Fatty Arbuckle's case, collecting fingerprint samples and placing hair follicles atop glass slides, while he was also preparing for William Hightower's trial, which would begin in a few weeks. Oscar was featured daily in headlines from both cases.

"The Arbuckle Trial," read one headline. "What Heinrich Saw Through His Microscope! Sherlock Holmes in Real Life."

The criminalist had now processed all the evidence and made a declaration—Arbuckle deserved to be in prison. Oscar despised the actor and the Hollywood excesses that he had represented. The press made much of the alleged crime, describing in detail how Virginia Rappe likely died under the star's large body. Oscar Heinrich was looking forward to trapping Fatty Arbuckle in court using forensic science.

"By the way the new drink down here in bootlegging circles is the 'Arbuckle Crush,'" Oscar crassly joked to Kaiser. "Fatty is guilty as hell of everything charged."

By mid-September, San Francisco's district attorney Matthew Brady had already spent several weeks diligently building his case, gathering statements from three showgirls at Fatty Arbuckle's party along with an affidavit from a nurse who had treated Virginia Rappe before she died. The woman who had kept a vigil by Rappe's side, Maude Delmont, would be his key witness, which was tricky because Delmont's affidavit was inconsistent. In fact, none of the witnesses seemed to totally corroborate one another, and yet their summary of the night's tragedy was similar—Fatty Arbuckle had followed the starlet into his private suite, assaulted her, and then joked as she lay dying.

"Arbuckle took hold of her and said, 'I have been trying to get you for five years,'" read Maude Delmont's affidavit.

Those quotes, printed in newspapers across the country, had sharp consequences—even before a jury was selected, one of the most respected actors in America was blacklisted by Hollywood. Many movie theaters banned his films, including one of the actor's biggest bookers. Theater mogul Harry Crandall owned eighteen elegant, upscale movie houses along the East Coast. Once enamored of the comedian, Crandall became repulsed by the vile details.

"The evidence adduced since the young actress' death constitutes one of the most repulsive crime stories ever printed," said Crandall.

Within weeks of Rappe's death, one of America's most beloved actors had vanished from movie screens.

District Attorney Brady was excited to begin the trial, a spectacle that would surely stoke his own positive public image. The forty-five-year-old was part prosecutor, part politician—an intelligent attorney with career ambition. Just one week after the now-infamous party, Brady quickly orchestrated two concurrent court proceedings, a coroner's inquest and a grand jury hearing. On September 12, the coroner's jurors sat inside an office in the Hall of Justice, waiting to hear medical evidence that would explain how Virginia Rappe died—and if Fatty Arbuckle was a killer.

A physician who had treated Rappe at the medical facility testified along with the two doctors who had performed the postmortem exam. They agreed that a ruptured bladder had caused her death, but there was no definitive evidence of violence. Two attending nurses described Rappe's complaints about chronic health issues, including abdominal pain. They testified that Rappe recalled few things about that night—her memory was hazy.

"She said that Arbuckle threw himself on her," testified Jean Jameson. "At other times she would say that she did not remember what happened after she got into the room."

Jameson testified that Rappe remembered having only three drinks. And the nurses declared that the actress never accused Arbuckle of hurting her, at least not in front of them. One nurse did say that Rappe made a small, private confession.

"The patient admitted to me that her relations with Arbuckle in the room had not been proper," testified Vera Cumberland. "She did not say whether her actions were voluntary or involuntary."

Coroner's inquests were frequently banal, clinical trials filled with medical jargon delivered by eggheads in white coats, so the panel suddenly became alert when the state's most intriguing witness, showgirl Maude Delmont, walked nervously toward the witness chair. The thirty-eight-year-old, swathed in black, peered down and fidgeted in her seat. Her answers were quiet, tentative, and sometimes lively, like when she was asked about her impression of Arbuckle at the party.

"I don't like fat men," Delmont replied.

The showgirl described a tipsy night of liquor binges, dancing, and singing to tunes spinning on the Victrola record player and then detailed what she told the police.

"I'm dying, I'm dying, he did it," cried the victim, according to Delmont.

The official in charge of the inquest, coroner Thomas Leland, was skeptical.

"How do you know what happened if you had so many drinks of whiskey?" asked Leland.

"My memory is always good," replied Delmont.

District attorney Matthew Brady was concerned about Delmont's vulnerability as a witness because, he admitted, her memory didn't seem to be clear. He worried that she wouldn't sway a grand jury to indict Arbuckle for murder, and Brady earnestly believed that Rappe was *murdered.* He made a strategic decision to exclude Delmont from testifying before the grand jury, which was meeting later that day.

That panel gathered in San Francisco's Hall of Justice to decide if there was enough evidence to indict the actor for murder. Those jurors were asked to consider the entire scope of evidence against Arbuckle, not just medical opinions. The jury sat for hours listening to the testimony of doctors, Hollywood players, and showgirls. Arbuckle, as expected, refused to answer any questions. But there was a hitch in the prosecutor's case, and the jury knew it. Matthew Brady's toughest obstacle was not Fatty Arbuckle's attorneys; it was a showgirl. And it wasn't Maude Delmont.

Zey Prevon slipped into the witness chair near the grand jury and vowed to tell the truth. She was asked about her police affidavit, the statement that confirmed Delmont's story and then convinced the district attorney to pursue murder charges. Prevon told police that Rappe cried out, "He did it," but now sitting before a grand jury, Prevon denied it all, and she declined to testify. She also refused to sign another formal police statement. The prosecutor was furious because his shaky case against Arbuckle was suddenly weakening even more.

"We have sent Miss Zey Pryvon [Prevon] home under surveillance," Brady told the press. "The girl changed her story completely before the grand jury."

Brady accused Fatty Arbuckle's team of witness tampering.

"I am convinced that undue influence and pressure of a sinister character had been brought to bear on her and other witnesses," he told the press.

The prosecutor spirited away Zey Prevon and Alice Blake into protective custody to guard their testimony. Prevon felt trapped inside a strange home, watched by police officers while Arbuckle's defense attorneys accused Brady of intimidation, threatening both witnesses with jail time. And now an inflamed media, infected with yellow journalism, fixated on Zey Prevon.

Newspaper journalists stalked her, snapping her photo in court and searching for tawdry details about her personal life. Reporters ridiculed her entertainment career, joking that her credentials were limited to disrobing for bathing suit newspaper ads. She feared the defense would belittle her on the stand, but the district attorney didn't trust her, either.

As Matthew Brady hurriedly gathered more evidence for the grand jury, he ordered Prevon and Blake to his office for a scolding. They must sign new statements, he told them, and they had to match Maude Delmont's accusations. Brady refused to allow a trio of showgirls to derail his case.

Prevon surrendered. She stepped back in the grand jury room and testified that yes, Rappe *had* directly accused Arbuckle of hurting her—Maude Delmont was right. Jurors listened, deliberated, and then handed Matthew Brady a partial victory by indicting the actor for involuntary manslaughter, a less serious charge than murder. Jurors concluded that Arbuckle hadn't planned for Rappe to die, but he was responsible—Zey Prevon had convinced them.

Brady returned to the coroner's inquest the next day and notched another win. The two physicians from Rappe's autopsy had concluded that she died from the inflammation of the peritoneum, a membrane in the inner abdominal wall—peritonitis caused by a rupture of the urinary bladder. They had also testified that "some force" caused her fatal ruptured bladder, like "finger pressure," but there was no way to be certain. The coroner's jury deliberated and, like the grand jury, concluded that the appropriate charge would be manslaughter.

And those jurors added one note of judgment—a condemnation of anyone associated with crime; the panel feared that San Francisco would become the "rendezvous of the debauchee [deviant] and the gangster." Now two separate juries had recommended Fatty Arbuckle be tried for manslaughter. He was now facing up to a decade in prison.

While he was in jail, the actor fretted over his unraveling reputation along with his safety because he was receiving death threats. His movie career was seemingly over. "Fatty Arbuckle Is Done," proclaimed the *Santa Ana Register.* Old friends turned against him, and he had to endure even more court hearings before his trial would begin in November.

On September 24, Arbuckle stood before a police judge in a preliminary hearing that could decide there had not been enough evidence for a trial. Zey Prevon was presented to a jury once again. She was clearly flustered, but she affirmed her revised story and added more details, like her description of Arbuckle's appearance as he emerged from the suite.

"He was fooling with his bathrobe, kind of tying," Prevon testified.

As Prevon watched the actor mill around the suite, she noted that the back of his pajamas looked wet. Prevon walked over to the bed as Rappe continued to moan—the sheets were also soaked. Jurors listened as she described undressing Rappe to make her more comfortable while Alice Blake, along with Maude Delmont, joined her as Arbuckle left the room.

"She said, 'I am dying, I am dying,'" testified Prevon. "'He hurt me.'"

The showgirl testified that Arbuckle, angry and drunk, stormed back into the room.

"He came over and said, 'If she can't stop her yelling I will throw her out of this window,'" testified Prevon. "She stopped then."

She described one of the most infamous, heinous scenes that would remain grafted to Fatty Arbuckle's character forever, a claim repeated from the earlier testimony of Rappe's manager. Both Prevon and Al Semnacker claimed that the actor had inserted a piece of ice into Rappe's vagina.

"'That will make her come to,'" Prevon testified that Arbuckle said. "Mrs. Delmont said, 'Oh don't do the'—shoved his hands away."

This was a new allegation, although decades later the story would morph into speculative lore about a vicious, horrific sexual assault, which sometimes involved a Coca-Cola bottle, a champagne bottle, or a broom handle. The prosecutor refused to repeat the allegation about the ice in court because it was simply not true, but the graphic details made years after Arbuckle died transformed him from legendary to notorious. Under aggressive cross-examination Rappe's manager eventually admitted that he might have been wrong about the ice.

"Are you aware that there is no medical evidence to back your claim?" asked Frank Dominguez, Arbuckle's attorney.

"I saw ice there," replied Semnacker, now squirming in his seat.

"But you did not see Mr. Arbuckle put it there," retorted Dominquez.

"No, probably not."

Now Arbuckle's defense attorney walked toward Zey Prevon, the state's most vulnerable witness, as Brady braced himself. Dominguez, a skilled interrogator, unleashed a litany of accusations. He demanded that she admit she had discussed her testimony with other witnesses, like showgirl Alice Blake, against the rules of the court. No, Prevon insisted, and Matthew Brady had *not* threatened her with a perjury charge.

Dominguez asked Prevon about Maude Delmont, the woman whose provocative statements had done so much damage to Arbuckle's defense. The defense attorney characterized Delmont as a pathetic witness who was so crooked and unseemly that Brady refused to let her testify. He accused Delmont of being an opportunistic harlot who became blindly drunk that night. Prevon told police that Delmont had consumed more than a dozen Scotch whiskys during the party.

"I don't know what she was drinking," Prevon told Dominguez from the stand. "She was drinking everything."

Arbuckle's attorneys pivoted to Virginia Rappe, the victim—an inebriated party girl who tried to tear off her own clothing.

"She was plain drunk at that time, wasn't she?" yelled Dominguez. "Acting like a hysterical woman at that time, wasn't she?"

"No, sir," insisted Prevon. "I thought she was sick."

Arbuckle's attorney blamed the victim and shamed the showgirls who were now witnesses. Alice Blake testified next. She said that Arbuckle *might* have been in the room when Rappe said, "I am dying, I am dying, he hurt me."

"You are crazy," Arbuckle screamed, according to Blake. "'Shut up or I will throw you out of the window.'"

Prevon's and Blake's stories seem to match, which was welcomed news for Matthew Brady. Finally, the district attorney called to the stand a chambermaid who heard a woman's screams from the suite that night—one of the maids who Oscar had refused to interview because she was "coo-coo."

"I heard a man's voice say, 'Shut up,'" testified Josephine Keza.

Arbuckle's attorney, fighting for his client's life, accused Maude Delmont of plotting with Rappe's manager, Al Semnacker, to extort money from the actor by using Rappe's torn clothing, which they had saved. The defense attorney discovered that Delmont had a string of criminal fraud charges, and police had accused her of attempting to extort money from another actor who had gotten her pregnant. There were rumors she was a madam. But inexplicably, Dominguez refused to call Delmont to the stand. He also decided not to allow Arbuckle to testify—two potential blunders.

When the testimony concluded, the judge ruled that there was not enough evidence to put Arbuckle on trial for murder; even worse for the state, the judge was incensed that Matthew Brady's lead witness, Maude Delmont, was not subpoenaed.

"Well, I will tell you one thing, Mr. District Attorney," declared Judge Sylvain Lazarus, "you are taking a chance on a motion to dismiss when you regulate your testimony so strictly that you are only putting in just what you consider sufficient."

Despite missing a key witness, Judge Lazarus ruled that Arbuckle *should* face a trial on manslaughter charges. The actor was to be released on $5,000 bail, and he would immediately return to his Los Angeles mansion to prepare for his trial in November. But first, as Arbuckle prepared to leave the defense table, the judge offered a parting thought.

"We are not trying Roscoe Arbuckle alone," said Judge Lazarus. "We are trying present-day morals, our present-day social conditions, our present-day looseness of thought and lack of social balance."

6.

Indignation: The Case of the Star's Fingerprints, Part II

It is a capital mistake to theorize before one has data. Insensibly one begins to twist facts to suit theories, instead of theories to suit facts.

—Arthur Conan Doyle,
A Study in Scarlet, 1887

The trial of the century opened on November 14, more than two months after Virginia Rappe's death. Arbuckle's attorney, Frank Dominguez, had been replaced by Gavin McNab, a well-known provocateur in criminal trials. McNab hired an investigator to probe Virginia Rappe's past in her hometown of Chicago, an aggressive strategy used to castigate the victim. He was particularly interested in rumors about several abortions.

In court, both sides quarreled over the recollections of witnesses, along with the cause of Rappe's death. The jury listened to torrents of information, mostly contradictory. Several physicians testified that the ruptured bladder could have been caused by violence, but perhaps it had happened spontaneously. And still absent from the stand was Maude Delmont.

Alice Blake's testimony added little to either side's argument, but Zey Prevon's recollections unsettled the case once again when she altered her original story. She refused to confirm that Rappe had accused Fatty Arbuckle of hurting her; she would admit only that the actress said, "I'm dying." In this new version, the actor had not assaulted Rappe with a piece of ice—he had seemed to hope it would help her.

"He put a piece of ice on her body and said: 'That will make her come to,'" testified Prevon.

The testimony proved to be a disaster for the prosecutor, who regretted his decision to trust a washed-up entertainer. And now each of the showgirls at the center of Fatty Arbuckle's case would slowly vanish from newspaper headlines.

"Call Heinrich," declared prosecutor Matthew Brady. Oscar walked across the courtroom, a large roll wrapped in thick paper underneath his arm. He placed it near the judge's desk as he slid into the uncomfortable wooden chair on November 23, 1921.

"Heinrich was humorless, cold, quiet, statistical, unflattering, as patient as Job," declared one reporter, "and it was clear that he cared nothing under heaven but the integrity of his theories."

Defense attorney Gavin McNab suspected that Oscar Heinrich would be the state's lead forensic expert—and so it was his job to lash the criminalist in court. Oscar had looked at the witness list for the defense. He snickered at one name: Chauncey McGovern, the handwriting expert from William Hightower's trial. Oscar was just complaining about McGovern once again to Kaiser, this time accusing his rival of plagiarizing someone else's invention in a magazine piece on composite photography. McGovern's arrogance peeved Oscar—he hoped to have a chance to publicly humble him. That day was still a few years away . . . but it would come. For now, he would have to contend with McGovern sneering at him from the sidelines, as he would not end up called to the stand.

McNab had earlier called Kate Brennan, a chambermaid at the

St. Francis, who testified that she had thoroughly cleaned Suite 1219 eleven days *before* Oscar Heinrich and his secretary had entered the rooms. The housekeeper became a sensation in the courtroom when she offered the jury a demonstration of her cleaning technique by picking up a dustcloth and polishing the woodwork of the courtroom. She had performed well under a blistering cross-examination in a preemptive strike from the defense to defame an arrogant forensic scientist and his dubious fingerprint evidence.

On the stand, Oscar unfurled the paper roll, revealing more than ten large photographs. The jurors listened and then squinted at his enormous pictures of the arches, loops, and whorls that formed Arbuckle's fingerprint pattern. Oscar detailed how he spent three days collecting evidence at the hotel, hauling his large microscope to the suite to examine the floor. He described his substantial discoveries, including "a large amount of dust, many specimens of women's hair, some phonograph needles, and a white feather." If the room had been cleaned, as the chambermaid claimed, then she had done a very poor job.

Oscar also testified that there were scratches on the door paneling, perhaps the result of a struggle. He explained how he used his high-powered microscope to match hair collected inside the suite with threads of Rappe's hair. But his most compelling evidence was one single photo, an image that he found of two hands superimposed—a clear sign to Oscar that Arbuckle had tried to prevent Rappe from leaving the room.

Gavin McNab, Arbuckle's defense attorney, listened closely before springing up when it came time for cross-examination.

"How do you know that among all the millions—I mean hundreds of millions—of people in this world," asked McNab, "there are not some who have finger patterns like those of Virginia Rappe and Roscoe Arbuckle?"

"I do not know that," answered Oscar. "There may be in the world some people whose fingerprints would be alike."

"Then all this is guesswork?" McNab replied.

"Not guesswork," Oscar replied. "But deduction gained through the law of averages, scientific experiment, and psychological knowledge."

Oscar explained that circumstantial evidence was like a collage that pointed to what really happened during a crime. If the handprints matched those of Rappe and Arbuckle, then it was *likely,* beyond a reasonable doubt, that they belonged to them and no one else. Before DNA testing was developed, analysis of a finger's ridges, spirals, and loops was considered the gold standard in criminal investigations. And fingerprinting would play a crucial role in this case.

Fingerprint patterns were noted by medical academics as early as the 1600s, but they were professors of anatomy, not scientists. In 1858, Sir William James Herschel, a British Civil Service officer in India, used handprints as a signature on contracts to identify natives in the Bengal region. The first known use of fingerprinting as identification in the United States came in 1882, when American geologist Gilbert Thompson in New Mexico used his own thumbprint on a document to prevent forgery.

In 1892, Sir Francis Galton, an English anthropologist, was able to establish that no two fingerprints were exactly the same by noting differences between individual samples—minutiae points found at the end of friction ridge and bifurcations (the spot where a ridge will split into two ridges). Fingerprinting would swiftly become an invaluable tool for criminal investigators, and "Galton's Points" became the foundation of the science of fingerprinting. Those points have since been used in automated computer programs to compare the patterns of two fingerprints in criminal cases.

In 1918, French criminalist Edmond Locard pioneered the science of poroscopy, the analysis of sweat pores on fingerprint ridges and surfaces, for individualization in criminal cases. Locard believed that if twelve specific points were identical on two different fingerprints, then investigators could declare that they belonged to the same suspect.

By 1921, the accuracy of fingerprint analysis was rarely refuted. But in 2009, researchers with the National Academy of Sciences questioned its accuracy in their report on the state of forensic science.

"Not all fingerprint evidence is equally good, because the true value of the evidence is determined by the quality of the latent fingerprint image," concluded the report. "A latent fingerprint that is badly smudged when found cannot be usefully saved, analyzed, or explained."

That meant that a latent (hidden) fingerprint found on the handle of a gun or a drinking glass would not be the same high quality as a print sample recorded by the Department of Motor Vehicles used for identification and security. And many times, even those sample prints have to be collected multiple times for accuracy. There is simply no quality control over fingerprint evidence.

There's also a severe lack of scientific validation, according to the NAS report. Fingerprint analysis is a technique that is vulnerable to incorrect interpretation because the analyst might be poorly trained or simply wrong, even while using a sophisticated computer program.

But in 1921, Oscar Heinrich was confident that he was right—Fatty Arbuckle had tried to stop Virginia Rappe from leaving his suite that night. The problem with that belief, we know now, is that Oscar was depending on flawed evidence.

The most anticipated witness of the actor's manslaughter trial took the stand on November 28, 1921, to testify in his own defense. Arbuckle explained his version of what happened that night, a simple narrative meant to transform his public image from sinner to savior. He testified that he had retreated to Suite 1219 that night to change his clothes, and as he locked the door behind him, he heard strange noises coming from the bathroom. When Arbuckle swung open the door, it hit Virginia Rappe, who was wet from vomit and writhing in pain on the floor. He described carrying her into the bedroom and laying her on the bed at her request; he had offered her two glasses of water, but she rolled around violently and fell off the bed.

"She turned over on her left side and started to groan," Arbuckle testified, "and I immediately went out of 1219 to find Mrs. Delmont."

Both Maude Delmont and Zey Prevon claimed that they were knocking and kicking at the door before he finally responded, but the actor refuted that story: "She was tearing on the sleeve of her dress, and she had one sleeve just hanging by a few shreds," he said. "And I says, 'All right, if you want that off, I will take it off for you.' And I pulled it off for her; then I went out of the room."

Arbuckle also confronted the most salacious, controversial accusation—the claim from witnesses that he sexually assaulted Rappe with a piece of ice. Arbuckle testified that he *did* in fact see a piece of ice lying atop her abdomen. When he picked it up and questioned Maude Delmont, she replied sharply.

"'Leave it here; I know how to take care of Virginia,'" Arbuckle recalled. "Mrs. Delmont told me to get out of the room and leave her alone, and I told Mrs. Delmont to shut up or I would throw her out of the window."

He had threatened to throw *Maude Delmont* out of the window, not Virginia Rappe, the actor explained. Arbuckle seemed composed and earnest, but he was also a canny actor. The jury was required to stay focused on his statements, not his apparent sincerity—that edict seemed difficult, considering Arbuckle's magnetism. The actor offered the jury a believable version of Rappe's illness. And now Gavin McNab hoped to discredit Oscar Heinrich's scientific evidence, the double handprints that supported his unequivocal conclusion that the actor was a killer.

"At any time of that day did your hand come in contact with her hand on the door?" asked McNab.

"No, sir," Arbuckle replied.

By the time he had stood up from the witness chair, Arbuckle had successfully revised the theme of the trial—he was innocent and being punished because of his celebrity status. After three weeks of testimony, both sides issued their closing arguments, passionate pleas for justice and retribution. Oscar's scientific findings were rubbish, proclaimed the defense, and there was no trustworthy evidence to convict Fatty Arbuckle. On December 2, the case was sent to the jury, and a panel of seven men and five women would decide the actor's fate.

Jurors, locked in their room, discussed the testimony, the scientific evidence, and Arbuckle's own story. They argued for forty-four hours . . . and then gave up. They were deadlocked ten to two for acquittal—the majority of the panel didn't believe the prosecutor. One of the two jurors had waffled for much of the deliberations, while the other was a steadfast holdout who refused to be swayed. And she credited Oscar Heinrich with convincing her that Arbuckle was guilty.

"It was the matter of fingerprints purely in the final analysis that decided me," said Helen Hubbard. "Arbuckle failed to convince me with his story absolutely. Therefore, I voted for conviction and no power in heaven or earth could change my fixed opinion."

The country's most celebrated criminalist had successfully forced a mistrial all by himself, but few people were pleased with the result. Matthew Brady, the district attorney, was outraged but resolute—Fatty Arbuckle was a lout who deserved to be jailed. The actor was crestfallen because he wasn't exonerated and his reputation was still dismal. There would be little rest for either side, because Fatty Arbuckle would face a second trial after Christmas.

Privately, away from the courthouse, Oscar Heinrich sulked. He felt misunderstood by the jury and concluded that the mistrial was a failure, a condemnation of his investigative skills.

"The ability of the defense to create a conflicting theory and support it with the testimony of Arbuckle himself aided and abetted by his lifelong training in acting gave them at least one leg to stand on," Oscar complained to Kaiser.

The criminalist had once blamed the modern woman for America's crime wave. It was her sex appeal and disregard for restraint that provoked young men to violence—a conclusion much of the country shared.

"Parents seem to think they must watch their sons and protect them from evil," Oscar said. "But I say, 'For God's sake observe your daughters, too, and try to cultivate in them a little of the restraint that young women had in the days when crime waves were unheard of.'"

Now Oscar cursed not just promiscuous women but the entire entertainment industry, blaming its deference to omnipotent stars with no self-control.

"The case of Arbuckle and its present status reflects the acceptance by the American people of a double standard in which the woman pays for every transgression," he wrote Kaiser, "whereas the man frequently if not always escapes."

While women had won full voting rights the year before, sexual assaults in America were vastly underreported; when survivors did respond to the police, many times they were blamed for being culpable. The popularity of adventurous flappers with their sexuality on display left men scared of false accusations, while women and girls continued to be sexualized.

Christian reformers hoped a ban on alcohol would help quell domestic violence. One goal of Prohibition was national purification, a return to the traditional sexual norms of the Victorian era. But often young women were cruelly chastised in court, like when judges punished underage girls for their role in statutory rape. Some prosecutors believed that girls were willing participants in incest, and their characters were impugned on the stand. While more women were attending college and earning careers, they were also habitually harassed on the street by "mashers"—men who aggressively "flirted" or catcalled.

"Our nation depends for its existence," Oscar wrote to Kaiser, "the inalienable right of the woman of every stage of society to choose her partner in every matter concerning her."

The jury was swayed by Arbuckle's star power, Oscar lamented, and he was incredulous that ten of the twelve jurors had dismissed his testimony. He likened Fatty Arbuckle to the doomed king Belshazzar in Babylon, who had held a feast using containers looted from a sacred Jewish temple. In the criminalist's version of the biblical tale, Oscar represented Daniel, the wise man who was able to read God's handwriting on the wall, which were phrases that revealed Belshazzar's blasphemy.

"Like Daniel in the days of Babylon I gave them the interpretation of the writing on the wall," he declared to Kaiser.

Oscar was certain there would be justice for Virginia Rappe, and he vowed to make sure of it during the second trial: "And I feel that just as Belshazzar who gave that party in Babylon died in the night, so will Roscoe Arbuckle die professionally and financially in the evening of this affair."

Christmas in 1921 was glorious for many Americans, at least those with a certain amount of affluence. There were pine trees or cedars for sale on wagons in cities, while families living in the countryside sent the men venturing onto farms with sharp axes. They hauled home the trees, soon to be heavily decorated on Christmas Eve with stringed popcorn, pinecones, red and green ropes, or homemade paper chains, while more traditional revelers (or those who couldn't afford electricity) used candles gently secured to the branches.

A lucky kid could sip a Coca-Cola and chew on a Baby Ruth candy bar for ten cents. Boys penned letters to Santa Claus, requesting toy Lionel trains or wind-up tin ships, while girls begged for walking and talking baby dolls on sale for $2.98.

On Christmas Day, wealthy families might have dined on oyster soup, roasted suckling pig, and diced turnips in hollandaise sauce followed by a dessert course featuring small cakes and nuts. But there was also a parallel Christmas in America, one where Santa Claus rapped on the doors of poverty-stricken children across the country, courtesy of the Salvation Army. The bearded volunteer hauled modest gifts of mittens or hats in his sack. The charity organization, founded in 1865 in London, also delivered free Christmas meals, including turkeys along with imported oranges for needy families across America. Those scenes reminded Oscar Heinrich of his own childhood Christmas holidays in the 1880s.

"When Santa Claus came to me instead of the usual red coat and white whiskers he generally had a red band on his hat with some gold lettering on it reading 'Salvation Army' and had a black mustache," he wrote Kaiser in a somber, reflective holiday letter at the end of 1921. "I never have been able to reconcile these two."

He was still ashamed by his father's financial failings, his willingness to abandon his family. Oscar lacked opportunities because of his father's weaknesses. He recalled one Christmas when, as an eight-year-old, he was gifted a present from his Sunday school.

"I, the son of the poor carpenter, got a red apple," Oscar recalled, "while the son of the rich jeweler got three pencil boxes, two oranges and a bag of popcorn. That was thirty-two years ago and the incident still has some effect on some of my social views."

Oscar vowed to never welcome a Santa Claus from the Salvation Army into his home again. His fixation on money strengthened especially during the holidays. The more anxiety he felt, the more he chronicled expenses in his domestic distribution reports. One week's worth of insurance, extra furniture, literature, carfare, meat, and tobacco had cost him $74.37. He filled out his charts weekly, sometimes daily. And he worried about how prosperous he might be in 1922.

That year's Christmas was almost unbearable because Oscar's testimony had hung the jury in Fatty Arbuckle's manslaughter trial—a horrid outcome, as far as he was concerned. And now he was being forced to feign yuletide spirit to his family upstairs instead of reviewing evidence in his laboratory below.

"I'll have to prime myself for a day filled with extravagant exclamations over the various things various members of my family will have received," he wrote to Kaiser, "and which to me appear to be utterly useless to them or anybody else."

He could hear eleven-year-old Theodore and seven-year-old Mortimer playing loudly upstairs. The boys were both spirited children, full of enthusiasm and endless questions for Papa. They had inherited many wonderful attributes from their mother, including her straight dark brown hair, attractive faces with dark eyes, and loads of energy—though it manifested itself differently in Marion. Oscar affectionately labeled his wife the "nervous-type."

"If I hurry Marion too much she is liable to lose a meal and is apt to go all to pieces," he joked to Kaiser. "This is particularly true if we are trying to make a train or something of that kind."

Oscar frequently felt obligated to shield his wife from his horrible cases, as well as their money troubles—not unusual for a husband in the 1920s who didn't believe his wife could offer constructive advice on their poor finances.

"We used to buy our Christmas presents and get them paid for about the following September," he told Kaiser. "It has been my privilege and pleasure to carefully refrain from concerning her with the problems of getting a living."

Oscar was becoming increasingly morose about the moral weight of his career, the public pressure, and its scrutiny. He carried a heavy weight—in one year, he sent eight men to San Quentin's gallows. An insecure perfectionist who routinely disassembled the stories of criminals, Oscar felt vulnerable to embarrassing public judgments, particularly in the press.

"I am sick and tired of sending men to jail and wish a change," he declared. "During the past few months I have managed to release several who were falsely accused but even so the entire problem involved the strife of life and the continued contemplation of the bitter side of its failures."

It may have seemed melodramatic, but after a decade of criminal investigations, Oscar was still not accustomed to the heavy burden he carried. Lives were spared or condemned based on his expertise, but sometimes his credentials were not enough to satisfy juries.

"I am not positive that I am doing yet that for which I was created," said Oscar gravely. "Life is a series of frustrations."

He fretted over another trial in January that would feature Arbuckle's haughty, expensive attorneys, so there was little joy that Christmas for Oscar Heinrich. But there was one exception, a gesture he had adopted as a nod to his childhood sorrow during the holidays. Once each of Marion's family gifts were mailed and the boys' toys were wrapped, Oscar drove down the hill to downtown Berkeley, where the Salvation Army collected its money, and he offered the volunteers a donation.

There was far less fanfare in the press for the retrial of Fatty Arbuckle in January 1922—no surprise witnesses and no fresh rumors purporting to be facts. Brady's case against Arbuckle was faltering because witnesses were now less certain and there was little new evidence. The press predicted a speedy acquittal. While Brady rearranged his strategy, Arbuckle's defense attorney plotted to humiliate Oscar Heinrich on the stand on January 27.

"The right hand of the man clutching the hand of the woman," the criminalist explained while pointing to large photos. "The position of the woman's hand on the door appears to have been due to pressure exerted by the man's hand."

No one in the court was shocked by the allegation at this point—it had been a centerpiece of the prosecutor's evidence in the first trial. But Arbuckle's defense attorneys had since developed a shrewd strategy by using an anecdote from a police officer assigned to guard the suite. The cop claimed that Oscar and his female assistant Salome Boyle had made a startling entrance when they first arrived to examine the rooms in September.

"Is it not a fact that you introduced yourself to the assistant manager of the hotel as Sherlock Holmes and your secretary, Miss Boyle, as Dr. Watson?" barked Gavin McNab. The audience snickered.

"Not that I remember," Oscar calmly replied.

He was slightly amused, because any glib comments he made about his Holmesian tendencies were quips, not delusions. McNab was determined to embarrass him, but then the defense attorney stepped across a very tenuous boundary.

McNab hinted at inappropriate behavior between Oscar and his assistant—it was enough to elicit a roar of laughter from the courtroom. The defense attorney insinuated that Oscar had locked the suite's doors to ensure privacy with his attractive young secretary.

"[I locked the doors] in order not to be disturbed, as I was being shadowed," Oscar explained.

He shifted in the chair. He was a devoted husband, a God-fearing man. The criminalist fumed because his career and his reputation depended not only on his skills but also on the public's belief in his integrity. McNab was trying to tarnish his reputation.

Oscar stayed composed, silently waiting for the next ridiculous allegation, and McNab offered it quickly. The defense attorney claimed that the two overlapping handprints on the suite's door did not belong to Rappe and Arbuckle but to Oscar and his assistant. The criminalist denied those charges, and he even offered his own handprint samples, along with Boyle's, as proof. McNab sneered. Oscar had been publicly dressed-down, while jurors stared at him from their nearby box. And now Fatty Arbuckle might be acquitted because of McNab's trap.

Oscar's stomach churned. But then he watched August Vollmer, his favorite cop, stride past him and slide into the witness chair. The criminalist listened to the police chief's confident, measured responses. Vollmer confirmed each of Oscar's theories about the handprints on the door—indeed, they belonged to Arbuckle and Rappe. At least Oscar could always count on August Vollmer.

His wife, Marion, didn't appear to be particularly concerned about him during the trial, but his elderly mother worried. By 1921, Albertine Heinrich (now Roxburgh) was remarried to a Scottish man and living in Eureka, California; she had written Oscar letters for years, always penned in her native German. Oscar had sent her money at least once a month for decades, an obligation the forty-year-old proudly accepted as a teenager but one that now placed a greater burden on his own struggling family. In fact, Oscar's personal finances were in such dire straits that vendors and lenders regularly hounded the criminalist, even threatening him if he refused to pay up.

"Permit me to say that this remittance is not being made to answer to your boorish threats," Oscar wrote to a building materials supplier. "You are probably aware by this time that I don't scare easily and I would suggest to you that your best interest lies in refraining from writing letters to me which don't read well after they have left your hands."

But Oscar refused to let bullies discourage him from supporting his ailing mother.

"Please don't feel backward about getting wood and coal or any other thing that you may need to make yourself comfortable," Oscar assured her. "I am perfectly ready to supply everything you need except a desire to put money away."

But money troubles were the least of his mother's concerns. While Oscar had been barraged by loads of publicity from Arbuckle's two trials, little of it was encouraging. She sent him a worried note about his safety—and he responded with assurances.

"Don't be worried about anybody you may think may be an enemy of mine," he told her. "I have no enemies except those few to whom perhaps I owe money. Don't be worried if people talk about me. That's merely advertising."

Oscar's confidence was peppered throughout letters to everyone in his life except one—his intellectual equal in scientific cases, John Boynton Kaiser. Arbuckle's two trials had alarmed Oscar and tested his faith in the judicial system. Kaiser rarely judged him and his decisions. He refused to harshly criticize the criminalist and offered only unwavering support and kindness. With Kaiser, Oscar could buckle . . . if only just for a moment.

The criminalist privately confided to the librarian that he felt utterly beaten, like a mule kicked by his callous owner for weeks. Fatty Arbuckle's defense team had insinuated that Oscar was arrogant despite his humble background, a useless scientist bent on convicting an innocent man, and a moral degenerate on top of it all.

"Can a spirit humbled by adversity be pompous and bombastic?" Oscar asked Kaiser. "Is someone who has been whipped a braggart? No!"

Oscar was gutted by the reaction of the jury, the cruelty of the press, and the venom from other forensic experts. The verbal flogging delivered by Arbuckle's defense attorney traumatized him, like so many incidents from his youth.

"In childhood and youth, I have been challenged, pilloried, stripped, flogged and crushed," he told Kaiser. "Crushed by those around me."

Oscar brooded as jurors began deliberations on February 1, 1922. After less than forty-eight hours of debates, the panel was deadlocked, just like the first jury. Two months of additional trial prep work had meant nothing, Oscar lamented.

Frustration turned to fury when the jury was polled—ten to two to *convict* for manslaughter. Oscar had been just two ballots away from sending Arbuckle to prison. The jurors offered some clear reasons. The defense had decided against putting Arbuckle back on the stand during the retrial, which was clearly a mistake. Gavin McNab was so confident in an acquittal, in fact, that he had refused to make closing arguments. The defense attorney's bravado alienated jurors, they revealed, so they punished Arbuckle. The jurors concluded that they didn't have enough information to make a uniform decision. Oscar was incensed.

"Mr. Arbuckle came within an ace of being convicted," he complained to a friend. "There was nothing more than a friend on the jury that saved him in this case."

On March 13, a third jury listened to essentially the same evidence, the same witnesses, and the same arguments. The defense changed tactics and called Fatty Arbuckle back to the stand and then issued a strong closing argument. Jurors were handed the case on April 12 at 5:08 p.m. They returned five minutes later with a stunning verdict, considering the case's history: not guilty.

"One ballot, no talk," said juror W. S. Van Cott.

The state had failed to convince this jury of Arbuckle's guilt.

"From the time the state's case was completed," said juror May Sharon, "I felt it wasn't enough."

Arbuckle was elated, thrilled the ordeal was finally over. And he was hopeful he could return to Hollywood films.

"If the public doesn't want me, then I'll take my medicine," said Arbuckle. "But, after the quick vindication I received I am sure the American people will be fair and just. I believe I am due for a comeback."

He might have been officially cleared, but his movie career remained ruined. Fatty Arbuckle's films had been banned for months, and movie executives, wary of the star's scandal, posted letters around movie studios in Los Angeles in June. The edicts demanded a return to morality in movie theaters—an embargo of any film material that wasn't wholesome. The studios had begun censoring Tinseltown.

A Hollywood censorship board was established that soon ruled that Arbuckle should never work in the entertainment industry again. Eventually the decision was reversed, though he would remain unofficially blacklisted.

About a week after the final trial, Oscar Heinrich turned forty-one years old. At age twenty-two, he had promised to intentionally reflect on his life on every birthday. Unfortunately for Oscar, many of those memories left him melancholy. His youth was filled with burdens, responsibility, and few rewards, even on his birthday.

"I never had a cake or even a reminder that someone was pleased about it until after I was married," he told Kaiser. "Nobody but myself ever seemed to pay any attention to it."

Oscar was comforted by the adoration of his wife and two sons; he was satisfied with his business and his professional success, but he was aging, and it felt unsettling. "I move more deliberately, I require more sleep," he told Kaiser. "Within the past few days my estimable wife chortled over the discovery of two gray hairs over my left ear."

Oscar valued his sons and coveted their education. He vowed to protect their futures, though he noted they were quickly developing into two very different types of boys. "Did I tell you that Theodore could sing the song 'The Old Swanee River' and a few other English songs in Latin?" Oscar wrote his mother. "Mortimer does not bring home such good reports. He's raising the devil in school all the time although he seems to have time enough left over to get most of his lessons."

The Arbuckle case had thrust Oscar Heinrich into the national spotlight, but it also damaged his credibility as an expert witness. And fingerprinting in America, still in its nascent stage, also came under the scrutiny of critics. Skeptical jurors couldn't understand if it was a credible science, and Oscar was certain that the jury had been seduced by a charismatic movie star, even if Arbuckle had been maligned in the press.

"If the entire episode results in a general elevation of the tone of moving pictures," he told Kaiser, "I shall feel satisfied that the prosecution's work was well done. This I think really will come to pass."

The aftermath of Fatty Arbuckle's ordeal in San Francisco, much of it caused by Oscar Heinrich, triggered wave after wave of disappointments and failures for the actor. His wife of seventeen years, Minta Durfee, divorced him in early 1925 because he was in love with another woman. Four months later, Arbuckle married his mistress, movie actress Doris Deane. But in 1928, after less than four years of marriage, she divorced him amid cheating allegations.

Arbuckle was buried in legal bills. He sold his famous cars, along with his Los Angeles mansion, which was worth $100,000. He wallowed in alcoholism. Studios declined to hire him because movie executives refused to risk a box office failure—all of this despite being categorically acquitted.

But Fatty Arbuckle appeared resilient. He was determined to stay relevant in Hollywood, only he found it prudent to adopt an alias—William B. Goodrich (the facetious moniker Will B. Good would stick with Arbuckle for the remainder of his career). One of his closest friends, actor Buster Keaton, hoped to help Arbuckle—but the comedian, once cheery and charming, was now a sad sack.

"It was a dismal experience to watch him," remembered Keaton. "Roscoe just was not funny anymore, was like some washed-up old performer who knew he was through and was just going through the motions because he had no alternative."

In early 1924, Arbuckle turned from big-screen actor to feature film director by directing *Sherlock Jr.* Now considered a silent film classic, the comedy starred Keaton as a shy movie theater projectionist and janitor who wanted desperately to be a detective.

During the film's opening scene, Keaton's character peered at a book titled *How to be a Detective* through a magnifying glass. He fantasized about becoming a professional sleuth who unraveled complicated plots.

Keaton's character, under Arbuckle's direction, was a twisted tribute to Oscar Heinrich, according to people on the set. The forensic scientist who helped ruin Arbuckle's career had actually inspired him creatively. The actor had watched Oscar closely in court as he pointed to large photographs of fingerprints. Fatty Arbuckle noted how Oscar Heinrich had absolutely mesmerized jurors.

7.

Double 13: The Case of the Great Train Heist

By a man's fingernails, by his coat-sleeve, by his boot, by his trouser knees, by the callosities of his forefinger and thumb, by his expression, by his shirt cuffs—by each of these things a man's calling is plainly revealed.

—Arthur Conan Doyle,
A Study in Scarlet, 1887

"We will play it on one card," the three brothers agreed. "If we win, we win. If we lose, we lose all."

The breeze felt cool on his neck. There was no place quite as peaceful in autumn as southern Oregon's forests of dark green pines. The swaths of color, yellow and orange on the maples and dogwoods, were divine for travelers wandering toward a campsite. It was certainly a bucolic, remote mountain wilderness.

As he stroked his forehead, his fingers slid on the unctuous, greasy tallow that was smeared on his skin. The stuff darkened his face. Hopefully he would be mistaken for a Mexican worker toiling on the railway. A putrid, smoky odor drifted around him. His shoes were covered with salt and sugar gunnysacks secured with rope and soaked in the dark, flammable chemical creosote—an attempt to mask his scent. He sprinkled black pepper from a one-pound Carnation tin can in the hopes that the pungent mixture of alkaloids and resin would repel tenacious bloodhounds on the hunt.

It was Thursday, October 11, 1923, when Roy DeAutremont (also spelled D'Autremont) and his two brothers squatted in prickly brush in the Siskiyou Mountains on the border of California and Oregon. They stared at the tunnel located on the steepest railway in the country, the line that traversed the summit of the mountain range just one mile south of its next stop in Siskiyou County. The Road of a Thousand Wonders line had carried thousands of travelers between Portland and Los Angeles via San Francisco over 1,300 miles through the Cascades, past snow-capped Mount Shasta and across thousands of acres of wheat, apples, walnuts, and roses near the Columbia River.

The handsome twenty-three-year-old tugged at his hat as the three brothers gazed toward the train tunnel, number 13. They were awaiting their first view of Southern Pacific Railroad Train No. 13, an express train—double 13. The brothers had heard unfounded rumors that the train, which the media later dubbed the Gold Special, was carrying up to half a million dollars in gold, all the motivation they needed to hatch a clever scheme to escape their troubling lives.

The steep grade of the Siskiyou Mountains required running the train in two sections, with a mail-express car leading the first half, followed by four baggage cars and three passenger cars. As it crested the steep grade of the summit, the engineer would be forced to slow down to test the brakes at the top of the pass by bringing the train to a near stop. The brothers planned to wait for the first three cars to pass through the south end of the tunnel before sneaking aboard and overpowering the crew.

Roy gripped his .45-caliber Colt automatic revolver and glanced at the gunnysacks. His younger brother Hugh, just nineteen years old, clutched a sawed-off shotgun filled with Ajax shells that contained buckshot for "long-range loads." A heavy weapon with powerful recoil, it was unlikely that anyone could survive an accurate shot. Roy's identical twin brother, Ray, sat at his side, his eyes fixed on the tracks as he listened for the rumbling and chugging sounds to grow louder from No. 13's engine.

A "blasting machine," a small wooden red box and plunger made by DuPont with a geared motor inside, was wrapped inside a pair of overalls. It was a device the brothers had stolen from a construction site near Oregon City. Pushing the plunger would turn the pinion gear on a magneto, a type of electrical generator, sending a current to the fifty pounds of wires attached to blasting caps on the sticks of dynamite inside a gunnysack lying nearby. Roy carried a .45-caliber Colt automatic pistol with its traceable serial numbers scratched off. They hauled along loads of ammunition.

The DeAutremont brothers were reared with guns in their hands during their childhood in the wilderness; their father had allowed them to load rifles as soon as they were strong enough to hold them. Roy had killed numerous rabbits before his tenth birthday, but small animals were his only victims. The brothers listened for the train's whistle, shrieking in the distance—they were jittery.

"What do you think about it, little lad?" Roy had asked his younger brother just the day before.

"The breaks are against us," replied Hugh gravely. "There is not much chance, but there is a chance."

They believed there was about $500,000 worth of gold and money on the train, cash and checks stuffed inside thousands of pieces of mail guarded by a U.S. Postal Service clerk. It seemed like a fortune to three brothers, who had spent much of their lives struggling. This would be their first train robbery—and their last.

The two men, dressed in black and donning Stetson hats and spurs, pointed their pistols at the engineer, forcing him off the train after it rolled to a stop. They dragged the frightened travelers from their seats and lined them up outside. A passenger dashed toward the tracks before being felled by a gunshot to his back.

In the annals of great American Western films, *The Great Train Robbery* claimed its place as one of the earliest movies in cinema, the first Western, and a genuine Hollywood blockbuster in 1903. In less than twelve minutes, the silent film told the tale of a gang of outlaws pursued by a sheriff's posse after a deadly train robbery. The bandits were menacing gunslingers who seized control over dozens of passengers before eventually dying in a blaze of gunfire in the woods. The story concluded with the leader of the gang firing his pistol point blank at the movie's audience, a startling cinematic technique repeated almost sixty years later during the opening sequence of the first James Bond film, and it's been used in each subsequent movie in the series.

Pulp magazines and dime novels in the 1920s glamourized the Old West by depicting real-life robbers like Jesse James and Bill Miner as protagonists, heroes who stole from the rich on trains or stagecoaches. Young men, including the DeAutremont brothers, marveled at Hollywood Westerns that starred daring bandits riding atop horses as they raced alongside trains and fan-fired at their enemies. The Robin Hood tale of struggling Americans robbing the wealthy in grand fashion was appealing to a generation that had survived World War I just five years earlier.

Train robberies were common during the 1800s in the American Old West. As the country expanded, bandits targeted slow-moving locomotives carrying cash and precious metals to large cities, but soon railway companies added massive safes with armed guards; they hired the infamous Pinkerton Detective Agency to track thieves. Mail robbery was especially lucrative because banks routinely shipped large amounts of cash and valuables by registered mail.

The brothers had read that gangs in the East earned millions of dollars by robbing mail carriages—and they were right. Between 1919 to 1921, mail robbers had stolen about $6 million, and the federal government was finally forced to act. In 1921, the Postmaster General pleaded with President Warren Harding to assign American marines as guards, and soon more than two thousand military personnel patrolled trains and government buildings, including post offices. The marines were trained to use deadly force to stop a heist, a disclosure that should have frightened any hopeful robber. But the lure of millions of dollars in cash and gold was too strong for the DeAutremont brothers and other gangs, who demanded their share of the country's growing wealth thanks to new leadership in the White House.

In August 1923, President Harding unexpectedly died of a heart attack, and Vice President Calvin Coolidge took office. After Harding's death, his legacy began to blacken amid revelations of political corruption within his cabinet and the president's multiple mistresses. His successor served as a sort of conservative father figure, a Republican leader known as "Silent Cal," who said little but made key policy decisions that shifted the economy. Calvin Coolidge favored tax cuts and limited government spending, and the economy quickly grew by 7 percent between 1922 and 1927. With new prosperity, more people earned better-paying jobs and overall crime slightly decreased.

Prohibition still hobbled the country, encouraging organized crime and holding the homicide rate at its highest level in American history. Despite the enormous economic boom, many Americans were still underemployed—still in need of a big, quick payout.

Roy DeAutremont looked over his nineteen-year-old brother, a slight and short teenager with a baby face and blond hair. The twins wondered if recruiting Hugh was wise—they would be responsible for their younger brother's life during a risky mission.

"Hugh, you see what is in front of you," warned Roy. "We won't think hard of you if you turn back."

"What the hell do you think I am?" Hugh yelled. "I am not turning back."

The DeAutremont brothers were improbable criminals, in a way, because of their strong family values. Roy and Ray were born in 1900 in Williamsburg, Ohio, while Hugh came along four years later. There were two other brothers, Lee and Verne, along with their parents, Paul and Bella—a boisterous, busy family of seven.

Their mother believed strongly in spirituality, so the five boys had faithfully squirmed in pews since they had been baptized into the Catholic Church as newborns. Roy regularly attended Mass and Sunday school and gave confession; he had enjoyed learning the tenets of the Bible. Roy and Ray had each made it midway through high school before dropping out; they were well-educated students who could read and write—two men devoted to their family and each other. Roy and Ray were inseparable, and though their bond with Hugh was strong, it was difficult to usurp a twin's lead position in his brother's life.

Their father, Paul DeAutremont, had spent much of their childhood hunting for work, so the family moved around frequently, all across the US, never quite settling down in one area. Instability bred domestic discord—Bella and Paul fought frequently, and eventually their marriage deteriorated.

"Their marriage life got to be worse and worse," said Roy. "It seemed like they absolutely couldn't get along. It made it so disagreeable for us boys that Ray and I decided to leave."

Paul relocated to Oregon without his wife or children to eventually open a barbershop, and soon the twins settled down with their father. Roy went to barber college, hoping to be like Paul, but instead he took a job at the Oregon State Hospital in Salem—a foretelling of his bleak future. Ray labored at a shipyard about sixty miles away and became an avid reader in his off time.

The three brothers were the antithesis of the desperadoes who starred in their favorite films. But they subscribed to a belief that eventually poisoned all three of them, transforming them into volatile and dangerous criminals. The DeAutremonts firmly believed that the government had betrayed one of them.

At age eighteen, Ray became a member of the Industrial Workers of the World (IWW), known as the Wobblies, a labor union that was stocked with indigent working-class men fueled by beliefs of injustice. Police arrested Ray and thousands of other Wobblies under a new legal statute called criminal syndicalism, which made it illegal to commit crimes, sabotage, or violence to support industrial or political reform in the United States. Even freedom of speech was repressed. Between 1917 and 1920, during America's post–World War I era, twenty-two states and territories enacted these anti-labor statutes, laws that were created to punish left-wing idealists and union organizers.

Police arrested Ray in 1919, and he was sentenced to one year in Washington State Reformatory in Monroe. He had been a compassionate and hard-working boy, but in prison he devolved into an angry, embittered man filled with vitriol toward anyone with authority. When Ray was released in 1920, Roy tried to assuage him by begging his twin brother to return to his Catholic faith.

"I did not know him, he was so changed," said Roy. "I think that he is crazy. He tells me the religion that I had been raised to believe is all bosh. But because Ray was my brother, my twin brother, I took what he said with silence and a heavy heart."

Ray spent twelve months behind bars, seething over his wayward life, his misfortunes at the hands of the wealthy. Soon after his release he presented Roy with a proposal: they would commit a robbery, just one heist to earn enough money to last them a lifetime. They enlisted younger brother Hugh, and the men spent the next three years plotting the perfect train robbery.

But there were hitches—they had a difficult time settling on a plan, and they oftentimes considered backing out. Roy even suggested buying a comfortable little homestead for their entire family and making due with sporadic employment, embracing an honest living.

"We knew it would mean we couldn't help mother and dad and

the brothers," said Roy. "We were sick of life, tired of it all, and we didn't care."

Ray and Roy ventured into logging camps searching for steady work; the two brothers were handsome and charming, natty dressers with manicured nails. But they were also slim and short—all less than five feet seven and each weighing about 130 pounds. They didn't have ideal builds for fledgling lumberjacks.

"I have come within an ace of getting killed time and time again in the logging camp," said Roy. "The work was too hard for me."

But he was determined. He would never go back to that type of dangerous job. So how could a crafty young man make a living in the rural northwest? Roy had an idea.

The ground of their campsite, just two miles from the tunnel, was littered with empty bullet casings, more than sixty spent .45-caliber shells. Roy squinted at targets and practiced firing his Colt pistol. They pulled their triggers quickly. They trained at aiming from the hip, but they spent most of their time becoming more comfortable with "snap shooting," a technique used when a hunter quickly points and fires at a target in close range without careful aim. They never bothered practicing much with the sights, the devices on guns used to make long-distance aim more accurate for snipers. This train heist, the brothers suspected, would likely end with a bloody, close-range gunfight.

They spent weeks traversing the countryside near the Siskiyou tunnel using a map of the state of Oregon they had bought from a bookstore. A compass guided them to different camping spots. They stored more guns, clothes, ammunition, and food inside a small abandoned wooden cabin at Mount Crest. They hid a cache of supplies at a different location, including a metal probe for digging bullets from their flesh and blood-stopping powder for healing their wounds. Roy, Ray, and Hugh burned most of their belongings—gun casings, toothbrushes, fuse wires, even cooking utensils—hoping to erase all incriminating evidence.

During a reconnaissance mission, Roy eyed a group of men repairing the track on the south side of the tunnel, and it almost deterred them. But they were determined to never return to one of those dangerous lumber camps. And Roy had a girlfriend back in Oregon, a young woman he hoped to marry someday, if they escaped from this robbery alive. He had even bought several double indemnity life insurance policies on himself, so she would receive at least $30,000 if he was killed.

The DeAutremonts were bright men, but their plan seemed to be bedeviled from the start. Several days before the day the train was due to pull through Tunnel 13, the youngest brother announced that he would drive their Nash touring car almost two hundred miles north to Eugene to visit with their father one last time. Hugh planned to tell Paul DeAutremont that he and the twins would report soon to a logging company for work after going on a camping adventure.

But within the first few hours, Hugh plowed into a cow in the Siskiyou Mountains and wrecked the car. He had to spend several days in Ashland while it was being repaired. Hugh left the car with their father in Eugene and returned by train, but then he was stopped on the station's platform in Ashland and questioned by a special agent assigned to keep an eye out for illegal activity. Hugh avoided trouble with the investigator, but he was forced to walk almost twenty miles back from Ashland. All this should have been a signal—a good reason to just quit. But the brothers pushed through.

Roy rubbed his knee as he sat on the hill. The twins had their own troubles after Hugh left for Oregon. The pair decided to sneak down to the tunnel one night for a closer look. It was dark, and after they poked around they heard a freight train chugging toward the tunnel. The brothers hid, intending to hop aboard and ride back toward their cabin. Roy scrambled onto the platform as he heard the clanking of the train wheels growing louder.

"I hit the platform with my knee," said Roy, "and it just about paralyzed me."

He screamed and landed on the wooden planks, unable to catch up to the train. Ray glanced back and hopped off. It was useless—the twins would have to hobble to their cabin two miles away, "discouraged and depressed, ready for anything."

While Ray and Roy waited for Hugh to return, they reflected on their plan, now only two days away. Roy wondered if they were foolish—perhaps they should just move ahead with his suggestion to buy some property and live a simple life. But Roy lamented the certain loss of his young girlfriend, and the men had hoped to help their parents and brothers. The train robbery had to be done, they decided.

Hugh finally returned from Ashland by foot, hours late and almost broke. He was exhausted from the luckless trip, while Roy was still healing an injured knee. The twins glanced at each other and asked Hugh earnestly one final time to reconsider.

"Hugh, you don't know what you are going into," Roy insisted. "If we fail in this, it means you are a dead man."

Hugh was steadfast. "Boys, I don't give a damn. I will take my chances."

They popped open their pocket watches—it was twelve p.m. on October 11, and No. 13 was supposed to arrive around one p.m.

"It was pretty near always on time," said Roy. "We went around, Hugh and me, and left Ray standing at the south end of the tunnel."

Ray carried a sawed-off shotgun stuffed with buckshot; he was agitated as he chain-smoked cigarettes, leaning against the concrete wall. Hidden close to the mouth of the tunnel was the bag stuffed with dynamite along with the blasting machine, that fire engine red DuPont detonator. Roy and Hugh eyed the Southern Pacific train as it pulled through the north portal of the tunnel and began to slow. Engineer Sid Bates gently applied the brake.

"Hugh, go out," Roy yelled.

The brothers began to run as Hugh scrambled onto the blind baggage carriage, an open car behind the tinder. The engineer spotted Hugh and quickly pulled the throttle wide-open, increasing the locomotive's speed. Roy struggled as he ran alongside—his knee still wasn't quite right. He was frantic as he dropped the .45-caliber Colt pistol with all eight cartridges still loaded.

"The worst I was scared on the whole job was right then when I thought I was going to miss that train," said Roy. "I was running to beat hell and losing ground."

Hugh climbed down the car's steps while Roy pumped his arms, hoping to catch up to the train as it picked up speed. Hugh turned around, stretched his leg back, and offered his brother a foot. Roy lunged forward and grabbed Hugh's boot, crawling toward his leg, hurling his body toward the steps and onto the carriage. Roy was convinced that no one had seen them as they snuck toward the engine cab in front of the train. Roy whispered to Hugh to give the engineer his orders, while Roy would handle twenty-three-year-old fireman Marvin Seng, the youngest member of the train's crew.

"Stop your train with the engine cab just clear of the tunnel," Ray yelled as Bates adjusted the throttle. "If you fail to do so the fireman will take your place because you will be dead."

Roy turned to the fireman, who was now alarmed by the pair of young men with dark, greased faces, one armed with a revolver.

"If the engineer fails to stop the train with the cab just clear of the tunnel," Roy declared, "you are to take his place because he will be dead."

As Bates peered through his round, wire-rimmed glasses at the brothers, Roy became suddenly agitated.

"I told him to keep his eyes off of me," he said. "The engineer acted like he thought it was a joke. He could see we were young."

The engineer quickly applied the brake, and No. 13 slowly screeched to a stop. The engine cab poked out from the mouth at the south end, with the remainder of the train (including the precious mail carriage, containing the loot the boys wanted to grab) and its passengers trapped inside the three-thousand-foot-long tunnel. Hugh pointed his revolver at the fireman and the engineer and ordered them to step off the train.

Ray watched steam fill the tunnel and realized that their plan had actually worked—the brothers had hijacked No. 13. Roy was eyeing the mail carriage, their main target, when he noticed a man's head pop out of the open side door. U.S. Postal Services mail clerk Elvyn Dougherty spotted Ray holding a shotgun and, in a panic, retreated. Ray lifted his gun, quickly aimed, and fired, missing the clerk. Dougherty stuck out his head again as Ray gathered the suitcase with dynamite sticks inside. The clerk slammed the door shut and barred it, locking himself inside the mail carriage to protect its cargo. Roy scurried over to Ray—now they were both standing in front of the mail car. Ray handed Roy the suitcase with orders to place it, along with some extra sticks, at the mouth of the tunnel near the train car.

The twins had agreed earlier that Ray would run up the hill, where the detonator lay hidden, but in his panic, Roy made it up the hill first and quickly pushed the plunger. There was a tremendous explosion, a concussion that shook the tunnel and rocked each of the three passenger cars. Roy had miscalculated the potency of their dynamite.

The front of the train ruptured, and its wheels lurched off the tracks, ruining their plan to disconnect the mail car from the remainder of the train. Worse, the dynamite turned the mail car into an incinerator by overturning a coal-burning stove inside, setting the room on fire. The mail clerk, Elvyn Dougherty, burned to death at his post, leaving behind a wife and young son. The thirty-five-year-old's charred spine lay exposed in the rubble hours later.

"I killed the mail clerk," Roy admitted.

The entire train trembled—glass shattered around the passengers.

One man in the smoking car cradled his bloody head. Glass shards sliced the artery in another man's leg. The passengers assumed that the engine had exploded and were alarmed by the smoke and putrid smell.

The tunnel filled with black smoke, steam, and fumes. Flashlights were useless—the brothers couldn't see a thing. Roy crawled on the ground, desperately hoping to find Ray and Hugh. He could hear screaming. They wanted to uncouple the mail car, but now the carriage was smoldering, and the steel was scalding. Roy had grabbed the two metal pieces clasped together and was trying to pull them apart when something scared him. The train's brakeman, dressed in dark overalls, jogged toward the train's left side and waved a lantern aglow with a red light. Roy pulled out a gun, yelling to Charles "Coyle" Johnson to put up his hands.

"I told him that his life was in greater danger than it had ever been before," said Roy.

Sensing the robber's agitation, the thirty-six-year-old agreed to help. Johnson disconnected the steam hose, but when he grabbed the lever that controlled the coupling, the brakeman struggled to release it. It was no use.

"He said to uncouple it you had to pull ahead with the engine while you pulled that up," Roy said.

Roy didn't trust Johnson, but he allowed him to hop on the train—arms up in surrender—to tell his brothers that the engineer should move the train forward. Johnson jumped on the carriage, but Roy didn't notice whether he had actually put his hands in the air.

Roy flinched when he heard the blasts—the crack of a shotgun and a pistol shot. He listened. The brakeman was dead, he was certain. He would later learn that, as Johnson lay dying from a bullet to the stomach, he muttered, "That other fellow said to pull the thing ahead."

"I think he forgot to put his hands up in the air," said Roy. "They thought he had killed me."

The engine still hadn't moved, so Roy ran toward the mail carriage and stood at the car's opening. It was bewildering—smoke and steam everywhere encircling twisted, sharp metal that was once something useful but was now indiscernible. The incredible heat contorted the walls of the carriage while the paint on the exterior was peeling. The stench of sulfur was overwhelming. The tunnel was also severely damaged—the explosion broke and charred the interior timber posts.

Roy dropped to his hands and knees and began scrounging for money, anything they could steal that would redeem this dismal mistake. Ray, Hugh, and the engineer tried to turn the train by pushing the throttle to make it move. It was futile—the huge charge of dynamite had knocked at least one section from the tracks. Now all three DeAutremont brothers were glaring at Sid Bates as he vainly worked the throttle.

"The engineer wasn't trying to pull that mail car," said Roy. "We all thought so."

The brothers yelled at Bates and Seng, the fireman, to pull the mail car out of the tunnel *now*. The steam was thickening, and the black smoke burned their eyes. The hiss of engine was deafening. It was a chaotic scene for a peaceful place like the Siskiyou Mountains. Roy was furious at their luck, and as the fireman stood near the engine, the brothers whispered to one another. Roy nodded, snatched Ray's pistol, and aimed at Martin Seng.

"I shot him with his hands in the air," said Roy.

Seng fell to the ground after two shots, his striped cap still on his head. His eyes stayed open as blood dripped down his nose. Roy had killed two men: the mail clerk and the fireman. His brothers had killed the brakeman. Of the witnesses who could identify them, only Sid Bates, the engineer, was left alive. The fifty-year-old was the veteran of the train's crew, with a nearly thirty-year career with Southern Pacific. It was rumored that he would soon retire.

Roy faced the mail car, readying himself to go inside the blackness. It was too daunting—he refused to explore the carriage without someone there to protect him.

"We will have to see if we can get anything out of the car where it is," Roy yelled to Ray. "We can't get it out of the tunnel."

The twins stepped inside the car and choked on the smoke and fumes. They stumbled over bags and walked carefully around a hole in the floor. They both fretted over finding the burned corpse of mail clerk Elvyn Dougherty. Roy gagged.

"The mail was all just blown to hell," he said, "and all on fire."

He told Ray it would be at least an hour before the smoke would clear, and if they waited, a posse sent from nearby Ashland would certainly hunt them down and lynch them. The men watched Bates as the engineer sat peering out the window of the left-hand side of the cab. Ray called out to Hugh, standing in the engine room and holding his Colt .45.

"Bump him off and come on."

Within seconds Sid Bates was dead from a shotgun blast to the back of his head. Powder burns singed his skin. The brothers had just killed each of the four eyewitnesses—there was no one left to reliably identify them and no real evidence to link them.

The brothers were now murderers without money. If there had been any loot left to salvage in the mail carriage, it was hidden by the smoke and debris. They dropped the burlap sacks, along with a pair of leather gloves, a cap, and a small suitcase. They never bothered to collect the red detonator or the pair of greasy overalls still hidden in the thick weeds.

A .45-caliber Colt pistol was lying inside the tunnel, though it would be impossible to trace without a legible serial number. Without hope of gaining any earnings from their ill-fated train robbery, the DeAutremont brothers did the only thing they could—they ran. As the train creaked from the jolt of a sensational blast, the three hapless robbers disappeared into the beautiful thick of pines lining the mountains. By late that afternoon, the largest manhunt in United States history began, a panicked search that unleashed thousands of volunteers, deputies, federal agents, and bloodhounds across the Pacific Northwest.

Edward Oscar Heinrich lingered over the collection of evidence spread carefully across his heavy wooden laboratory table. He gave each clue just a cursory glance because he knew that a simple pair of spectacles would never solve the mystery of the vanishing train robbers. Two special agents, including one with the federal government, milled around his Berkeley lab mid-morning on October 16, 1923. Local sheriff's deputies in Oregon were poorly equipped to process a murder scene, so Southern Pacific Railroad and the U.S. Postal Service, the employers of the four murdered men, dispatched their own investigators.

By the fifth day, the seasoned agents were simply stumped, left inert by a mystery with few clues. Confused passengers couldn't even agree on the number of suspects. A baggage clerk saw just two men running toward the front of the train. The agents spent days examining the evidence left behind, only to label each piece useless. There was almost no trace evidence, no viable hidden clues to be mined.

"Nothing doing," the local sheriff told reporters, "except the finding of a couple of leads that may work out into something."

Despite scant evidence, the sheriff's deputies quickly made an arrest in a nearby town, hauling a shocked suspect to jail who seemed to be tied to the few circumstantial clues at the scene. And now the agents were crowding Oscar's lab, insisting they needed the forensic scientist to prove the man's guilt.

Oscar Heinrich's business had matured well since his investigations into the cases of Father Heslin and Fatty Arbuckle. He had accepted an unprecedented number of clients, and he continued to garner international attention, but he still seemed dissatisfied. The shadow cast by Fatty Arbuckle's multiple trials haunted him, chasing him from case to case, because most juries still seemed dazed by his scientific evidence.

In 1922 and early 1923, Oscar Heinrich had worked on more than one hundred cases including forgeries, kidnappings, disputed wills, and, of course, violent crime. He investigated the murder of Anna

Wilkens in San Francisco, whose husband killed her and then blamed carjackers. The still-unsolved murder of Hollywood director and actor William Desmond Taylor had stymied the nation's top detectives, including Oscar.

While most criminal trials were settled in Oscar's favor, juries still mistrusted science and doubted the experts who espoused its credibility. A brilliant forensic expert, Oscar now believed, was crippled in most American courtrooms.

"The more learned the chemist, the more unskillfully would he drape words and clauses about an idea," Oscar complained to John Boynton Kaiser, "until it resembled pictorially and structurally the benzene nucleus and side chains of his chemical notebook."

When a scientist defended the merits of his own methodology on the stand, many jurors responded with dreamy, distracted gazes. That scene, repeated over and over again in courthouses during Oscar's career, was simply maddening to the byzantine mind of a persnickety forensic scientist.

Investigators stood over the bodies of the three railway workers, now lying on the ground outside the tunnel. The postal worker's body was still smoldering inside the charred mail carriage. Within hours the sleepy countryside of Northern California and southern Oregon came alive with posses from both states hunting for clues. Bloodhounds dragged deputies along rocky trails, but it was windy, so the dogs had a tough time.

"Bloodhounds today failed to pick up the scent of the desperadoes," reported the Woodland *Daily Democrat*. "Other hounds from Seattle, Washington, Salem, Oregon and Yreka are to be given trails later."

Hundreds of Oregon national guardsmen, peace officers, and volunteers fanned out across the rural county, dragging along sharp tools for hacking weeds and poking at the ground. Lawmen questioned townspeople about their alibis and cautioned families to latch their doors. There was already talk of lynching the men once they were found, but an Oregon State militia was dispatched to stop mob justice. Rumors circulated that the bandits were former railway workers, while witnesses reported seeing a car dash through Ashland about an hour after the murders. Southern Pacific Railroad immediately offered a reward of $2,500.

"Mail Car Dynamited by Bandits in Daring Raid," read one headline. "Bloodhounds and Posses Trailing Bandits Who Robbed Train and Murdered Crew," declared another. "Posses Scour Hills for Bandits."

Investigators were overwhelmed with fruitless leads and unhelpful tips, including analysis from a local clairvoyant.

"It was the boldest train robbery since the days of the Old West," read an Ohio newspaper.

Newspaper editors seized on every new detail about the failed train heist, a tragedy transformed in print copy to a remarkable, bloody nightmare that hijacked high-society newspapers and low-level rags. Headlines publicizing those unidentified "killers on the run" frightened American readers. The legend of "America's last great train robbery" was solidifying, no matter that the train had not *actually* been robbed.

Nearby the Siskiyou tunnel, investigators pulled back weeds on the small area just above the tracks, revealing a blasting machine with two batteries attached lying close to a pair of greasy blue denim overalls. Creosote-soaked gunnysack shoe covers lay nearby. A deputy discovered a revolver in the tunnel, a government-issued .45-caliber Colt. Investigators crawled on the ground and measured three sets of footprints while officers rounded up and released local troublemakers spotted near the railroad yards.

The county had no lack of good-for-nothings to interrogate. Police briefly arrested a twenty-two-year-old ex-convict but soon released him when he provided an alibi. They held two drug addicts, but soon they were also sent on their way. Three hunters admitted to being near the Oregon and California border, but they were also cleared. Investigators winnowed their list to just one suspect—a man now sitting in a local jail thanks to his filthy fingernails.

On the hill near the tracks, a local sheriff's deputy squatted down near the detonator and examined the batteries. Perhaps they came from a mechanic's repair shop? The overalls were clearly smeared with grease in several places. Deputies raced to a repair shop in a nearby town, hoping to find a suspect inside; as they carefully approached an employee, he explained that he was the only mechanic on duty.

He denied ever owning the batteries, but his face and his fingernails, the deputies noticed, were covered in grease. He wiped off the grime as he insisted that he had nothing to do with the train robbery, but he had no verifiable alibi. As investigators eyed the mechanic, his denials grew louder. They needed more proof to arrest him, so one deputy pulled out their best evidence, the overalls draped over the detonator. He demanded the mechanic slip the pair over his clothes.

Surrounded by armed men, the mechanic quickly agreed and shoved both legs inside. The overalls fit—not perfectly, but reasonably enough. The mechanic argued that they had no real proof, but it was too late, and soon he was pacing inside a small cell, waiting to see if deputies would issue him the third degree.

The mechanic was lying, insisted the investigators, a butcher no better than Jesse James and his bandits on horseback.

The special agents spent days in the interrogation room with no results, and now they needed Oscar Heinrich to glean anything he could from the items. The criminalist began by examining a .45-caliber pistol, a brown canvas knapsack, burlap shoe coverings, and a pair of blue denim bibbed overalls. The agents explained that as many as twenty other experts had investigated the clues, and other than mechanic's grease, there were no important discoveries. As his assistant escorted the investigators to the door, Oscar turned to his desk.

He fingered the cloth, felt its stiffness, and then smelled it. He held up the overalls to a wooden door and gently pushed six pins into the denim fabric, spreading it out as it suspended just above the floor.

He leaned the door against his bookcase, which had been stuffed with references sent by John Boynton Kaiser that included *Catalysis in Organic Chemistry* and *The Properties of Electrical Conducting Systems.* He reached over to his desk for two measuring tapes, stretching them horizontally and vertically before jotting down numbers, details about the length and width: five feet by twenty inches. He chronicled each experiment with his camera.

Oscar discovered the tag from the manufacturer—"United Garment Workers of America"—and then noticed the bottom of the turned-up cuffs of the legs. He asked his lab assistant to fetch a special pair of worn leather boots with thick laces, a set he had saved for years. He placed them on storage boxes underneath the overall's cuffs, which hung just five inches above, and made more notes.

"The overalls were quite new and had not been laundered," he noted.

The left-hand pockets were used more frequently, and the suspenders were handled almost exclusively from the left-hand side.

"Suspenders on left fastened and unfastened habitually," he scribbled on a small slip of brown notepaper, "by oily condition of inner edge of bib on left side and oil on buttons; also oil and wear on suspender end and buckle—absence thereof on right side."

He frequently jotted notes on scraps of paper he found on his desk or at hotels, a cost-saving measure he had adopted years earlier. It may seem haphazard to store such important clues on discarded paper, but he secured each note, synthesized the details, and produced a thorough report for every case. Using his magnifying glass, he spotted hairs caught in the buckles of the overalls, slid them under his microscope, and enlarged them. They were medium to light brown in color.

"Every individual, particularly a man, accumulates dust and dirt typical of his occupation, at various points about his clothing, particularly the pockets," he told another college professor. "Careful search of a suit of clothes also will almost invariably reveal head and body hair."

Oscar's gaze traveled up the garment, and he stared at the engine oil on the left pocket, evidence that had convinced federal agents that a local mechanic was the killer. Oscar scraped off some of the dark sticky goo, spread it carefully on a glass slide, and placed it underneath his microscope. He adjusted the oculars and rotated the magnification dial. It wasn't grease, he was sure, because he didn't spot any of its standard components: mineral oil, vegetable oil, lime, or emulsified soap. He dripped a reagent on the slide and watched the chemical reaction—the goo was a purely organic substance.

With his pencil Oscar made the most important note of the case, a scribble on the back of an old envelope that would save the mechanic's life: "Pitch—not oil on left pocket." The grease on the overalls came from a tree, not a vehicle. And soon Oscar would determine that the pitch actually came from a Douglas fir, a tree found in western Oregon, the same type of naturally occurring, sticky resin used to caulk the seams of wooden sailing ships for centuries. Oscar turned out the pockets of the overalls carefully as little chips, caught in the fabric, reflected the light of his small flashlight.

"No larger than the size of half a pea," he wrote. "The pockets carried tiny chips, earth debris, and botanical debris peculiar to the western Washington and western Oregon forests."

The suspect lived in the western part of Oregon, he concluded. Oscar used a small electric suction device, like a mini vacuum cleaner, to collect any other trace evidence. He could later sort through the material and catalogue it. Also inside the pockets were small, hard white chips—fingernail clippings. The man trimmed his nails, a strange habit for a lumberjack.

"A man who carries a fingernail file," Oscar told the agents, "is usually fastidious not only so far as his nails are concerned, but about his general appearance."

Oscar understood human nature—a man's habits reflected his personality. This suspect took pride in his appearance, the criminalist was certain. Overalls typically included numerous pockets for storage of various workman tools, including a pencil, so Oscar knew he would need to search each one. He adjusted his glasses.

"Pencil pocket at the left side of the bib watch pockets, on the left front bib of the overalls," he wrote on a steno pad.

Oscar stood up and hunted for what seemed like a curious tool in a forensic lab: a crochet hook. He suspected that the federal agents had not actually searched all the pockets—and this was a small one. He delicately slipped the hook inside the pencil pocket and fished around, occasionally removing it to check for progress.

"I aim to return evidence as intact as it is when it first comes into my laboratory," he explained to investigators.

After several minutes of searching, something in the inside seam became caught, and he carefully removed a slip of paper "about the size of a cigarette paper." He held it in his hand—it was stiff, a tiny ball. He needed to unravel it very slowly; otherwise it would disintegrate. There was the slight clinking of glass from beakers as Oscar brushed a chemical over the packet, hydrating the paper before unrolling it and then carefully ironing it under a low heat. The words and numbers from the clue emerged: September 14, 1923, #2361. It was a U.S. Postal registered mail receipt from Eugene, Oregon—a traceable bit of evidence that might reveal its owner. Next Oscar examined the Colt .45-caliber, which had contained loads of information for the right detective.

"Pink stain in grip—toothpaste. White stain—shaving soap."

Much of the serial number was scratched off, leaving just three numbers: "C _ _ _ 763." Oscar remembered something that few investigators knew—for years, firearms manufacturers had inscribed a second set of serial numbers *inside* their guns to help establish ownership. When Oscar dismantled the pistol, he found the duplicate serial number inside the gun:

"Colt secret number under firing pin," he noted.

The mystery serial number was C 130763, and now investigators might be able to trace who had purchased it.

Oscar Heinrich spent nine hours examining the evidence and then another nine hours inside his lab the next day. When he finished, Oscar dictated a letter to his assistant, which was addressed to the chief special

Left: August Heinrich and family *(Courtesy: Millard F. Kelly/UC Berkeley)*

Right: Heinrich and Marion on board *Conte Biancamano*, 1930 *(Courtesy: EO Heinrich/UC Berkeley)*

Heinrich looking through ocular *(Courtesy: EO Heinrich/UC Berkeley)*

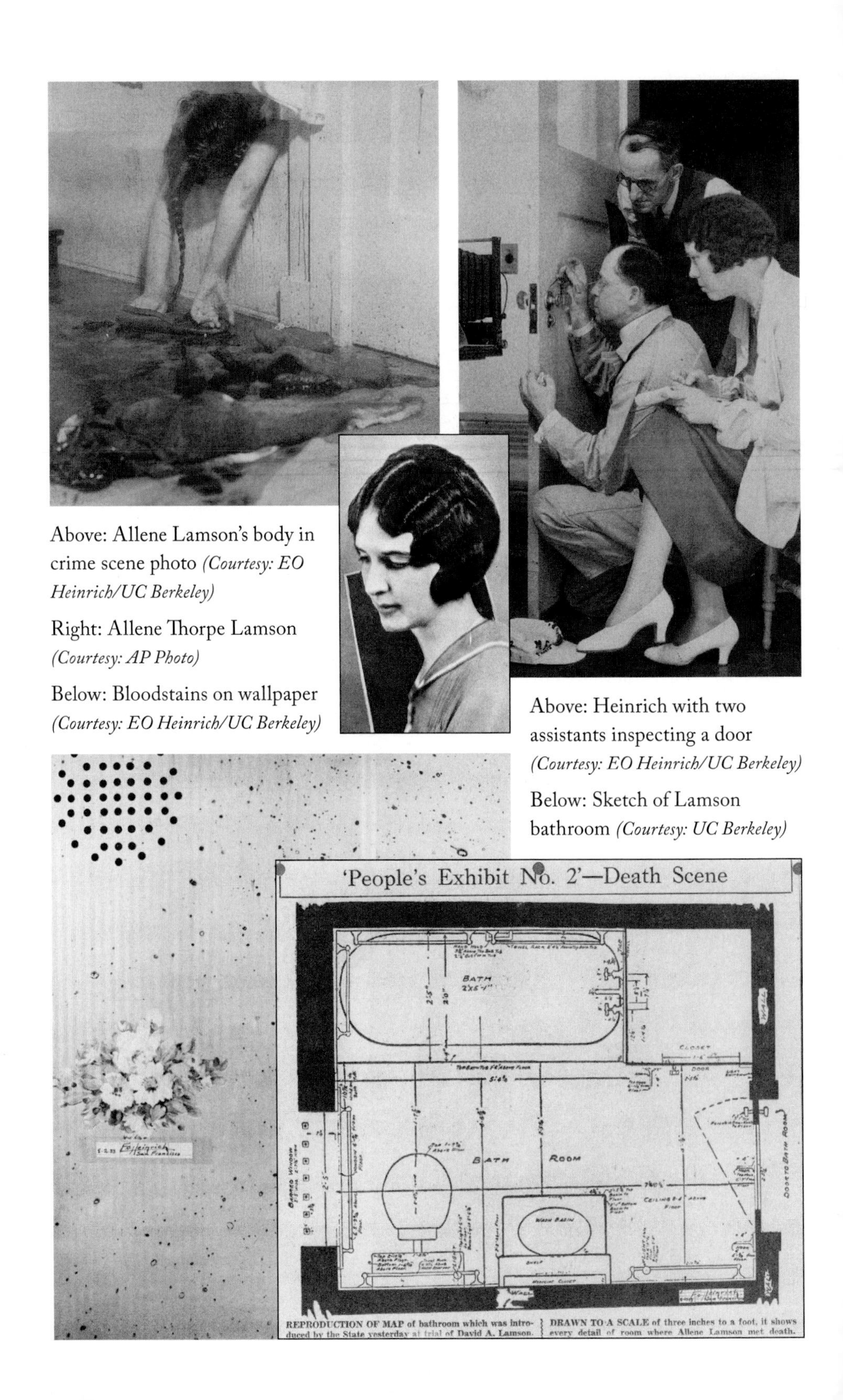

Above: Allene Lamson's body in crime scene photo *(Courtesy: EO Heinrich/UC Berkeley)*

Right: Allene Thorpe Lamson *(Courtesy: AP Photo)*

Below: Bloodstains on wallpaper *(Courtesy: EO Heinrich/UC Berkeley)*

Above: Heinrich with two assistants inspecting a door *(Courtesy: EO Heinrich/UC Berkeley)*

Below: Sketch of Lamson bathroom *(Courtesy: UC Berkeley)*

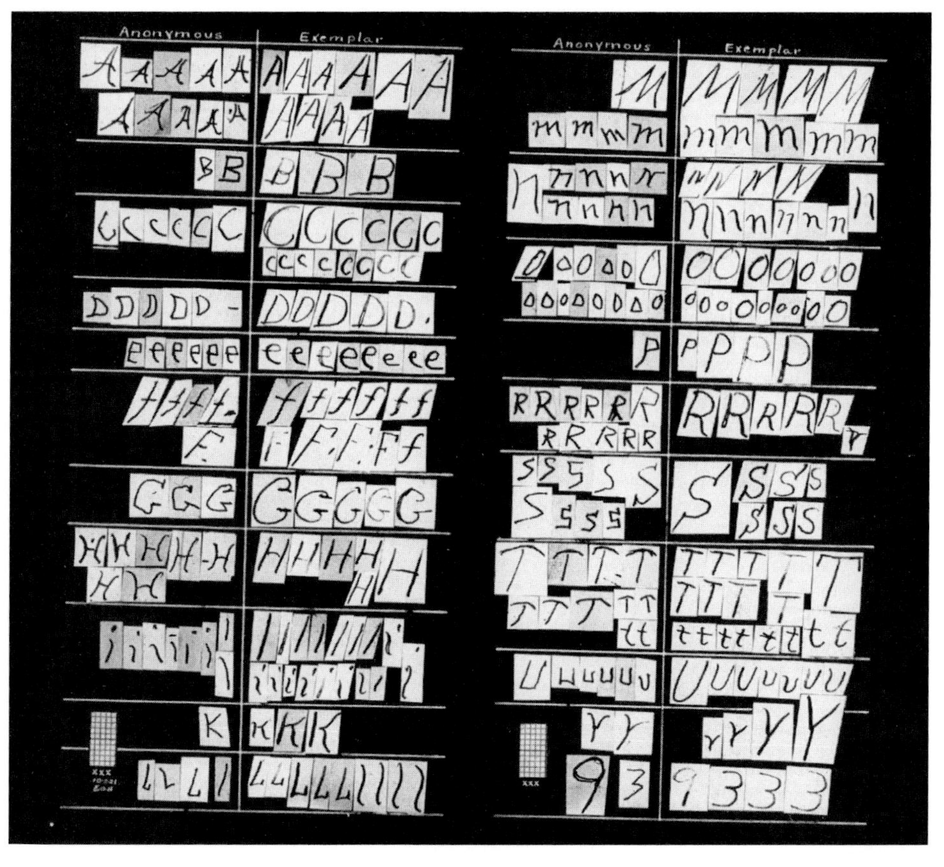

Letters for handwriting comparison in Hightower case *(Courtesy: EO Heinrich/UC Berkeley)*

Above: Close-up of knife with fiber *(Courtesy: EO Heinrich/ UC Berkeley)*

Left: William Hightower *(Courtesy:* The San Francisco Examiner*)*

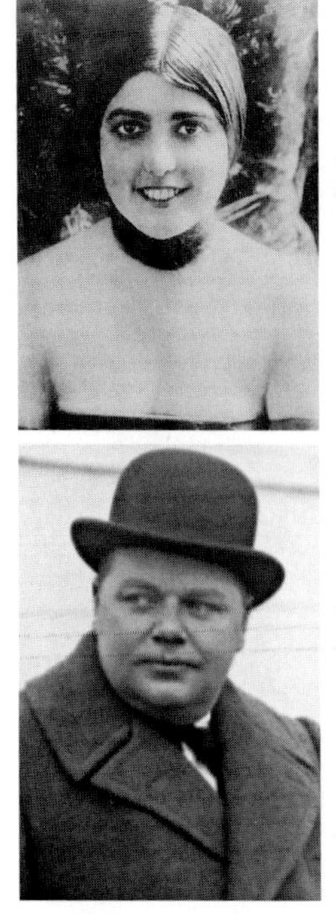

Left: Actress Virginia Rappe in 1921 and comedian Roscoe "Fatty" Arbuckle *(Courtesy: AP Photo/File)*

Right: Interior of Arbuckle suite *(Courtesy: EO Heinrich/UC Berkeley)*

Above: Close-up of handprints on door *(Courtesy: EO Heinrich/UC Berkeley)*

Right: Wide of handprints on door *(Courtesy: EO Heinrich/UC Berkeley)*

Left: Close-up of bullet markings in Martin Colwell case *(Courtesy: EO Heinrich/UC Berkeley)*

Right: Portrait of Bessie Ferguson *(Courtesy: UC Berkeley)*

Heinrich examining Ferguson's bones *(Courtesy: EO Heinrich/UC Berkeley)*

AMERICAN RAILWAY EXPRESS CO.

REWARD!

$14,400.00

Holdup of Southern Pacific Train No. 13, 1st Section, at Siskiyou, October 11, 1923

FOUR MEN KILLED

Reward of $2500.00 will be paid by the Southern Pacific Railroad Company, of $300.00 by the American Railway Express Company, and not to exceed $2,000.00 by the United States, for the arrest and conviction of each person implicated in the holdup.

At least three persons participated in the crime. Below are photographs and descriptions of three brothers who are believed to have been connected with the holdup and who should be arrested on sight and held incommunicado.

DESCRIPTIONS

[illegible]

All three men are loggers and may be found in logging camps working as choker setters, hook tenders or whistle punks. They have spoken of taking a long sea voyage and look-out should be kept for them attempting to ship at sea ports. They speak Spanish fluently and may attempt to cross the Mexican border. Formerly lived at Lakewood, New Mexico.

Photographs of Roy and Hugh were taken about a year ago; of Ray, taken in 1920.

Wire information, charges collect, to D. O'Connell, Chief Special Agent, Southern Pacific Railroad Company, Ashland, Oregon; C. E. Terrill, Sheriff, Jackson County, Jacksonville, Oregon, or C. Riddiford, Post Office Inspector in Charge, Spokane, Wash.

Postmasters will post and publish and distribute to Rural and Star Route Carriers, also see to it that all Peace Officers are supplied with these circulars.

C. RIDDIFORD,
Post Office Inspector in Charge,
Spokane, Washington

D. O'CONNELL,
Chief Special Agent, Southern Pacific R. R. Company
Ashland, Oregon

Case No. 57883-D

Ashland, Oregon, October 24, 1923.

Left: Ripped postal ticket and ticket for American Railways Express *(Courtesy: UC Berkeley)*

Right: Wanted poster for DeAutremont brothers, 1923 *(Courtesy: UC Berkeley)*

Left: Overalls pinned to wooden door, with measuring tape *(Courtesy: UC Berkeley)*

Right: Interior of train car taken from mail end *(Courtesy: UC Berkeley)*

Police reviewing the inside of Schwartz's laboratory *(Courtesy: Reynolds, Oakland PD/UC Berkeley)*

Above: Inside Charles Schwartz's laboratory *(Courtesy: EO Heinrich/UC Berkeley)*

Inset: Charles and Alice Schwartz *(Courtesy:* The San Francisco Examiner*)*

Lamson faces the court as his preliminary hearing opened at the San Jose, California, Hall of Justice, June 16, 1933 *(Courtesy: Associated Press)*

Heinrich at work in lab, examining a gun *(Courtesy: Herbert Cerwin/UC Berkeley)*

agent for Southern Pacific Railroad Company, Dan O'Connell. Oscar explained that he had discovered a wadded-up receipt for ten cents from the U.S. Postal Service that might be traced back to its purchaser. He had also recovered a hidden serial number inside the discarded .45-caliber Colt that could likely lead to the buyer. But O'Connell had made it clear that he needed specifics about the suspect, a way to describe him to the papers and eventually convict him—a profile. O'Connell and a federal agent with U.S. Postal Service traveled to Berkeley.

"A lumberjack," said Oscar. "The wearer and owner was a lumberjack employed in a fir or spruce logging camp. A white man not over five feet ten inches tall, probably shorter. Weight not over 165 pounds, probably less."

"Not so fast, professor," said the postal agent. "Do you mean to say that you found all of this out merely from examining those overalls?"

Oscar explained that he could estimate the weight of the owner because lumberjacks frequently bought their overalls in a larger size so they had room to shrink in the wash. These overalls appeared to be new.

"Hence the owner is undersized rather than an average sized man," he concluded.

The special shoes that his assistant retrieved earlier were logging shoes, and when he placed them underneath the overall's cuffs, they were the perfect length for a logger hoping to keep his clothes from being ruined while climbing trees.

"Lumbermen always wear their pants legs turned up in a deep cuff about halfway between the ankle and calf and outside the boot," Oscar noted.

He measured the distance from the cuffs' creases to the shoulder straps to estimate the wearer's height. He also knew that the owner's left shoulder was three-fourths of an inch higher than the right, based on how the straps were adjusted. The suspenders were handled exclusively from the left-hand side, and the pockets were most frequently used on the left side—though that didn't necessarily mean that the man was left-handed.

The suspect was Caucasian, according to the charts Oscar used to classify the two strands of hair. And while he was not able to use the hair to conclusively identify the suspect, his assertion of the ethnicity of the owner was scientifically valid. He had created a physical sketch of a killer—an incredibly accurate profile based on a single pair of overalls. Oscar's description tallied precisely with Roy DeAutremont—a brown-haired lumberjack with a slight build and height, a fastidious man who worked in western Oregon on spruce and fir trees.

"Putting these things together required no great stretch of the imagination to classify the occupation of the owner," he explained, with his normal trace of bravado.

Deputies released the mechanic from jail, and now special agents focused their search on a lumberjack from Oregon.

In the Superior Court at the state capitol Charlie Hicks had been battling for the widow all day long. Throughout the day his case had gone against him. Even a good dinner in the famous hostelry of the capitol failed to cheer him.

Oscar Heinrich stared at his typewriter, determining if he might tighten that section. Charlie Hicks had proven to be a difficult character to define.

He ordered mechanically; chewed belligerently; ate without tasting. Dinner over he wandered into the hotel lobby. He settled himself before the fireplace and truculently perused a newspaper.

"*The Black Kit Bag.* By E. O. Heinrich" was neatly printed on the first page, one of many detective stories he had written in his spare time, usually late at night after he finished his lab work.

He occupied himself with tearing his newspaper into odd shaped sheets, rolling each fragment into a tight paper wad and savagely flipping it at a blazing knot in the fireplace.

Oscar's passion for writing was first stirred during his teenage years; for the past thirteen years, Oscar had squirreled away fragments of paper with fictional story ideas along with longer ruminations on creative magazine articles and books. He kept a hidden folder of poems he had once penned for an old girlfriend in Tacoma before he met Marion—lines written with romantic intention.

Does your smile still dance out spritely,
From beneath your temples' steeps,
Skip about your eyes a moment,
Ripple down across your cheeks,
Lit with dancing beams of love light.

The verses seemed like a startling contrast to his tightly wound public persona. Writing was a venue for Oscar to spill secrets on paper that only he would review later in life, perhaps during wistful times when he thought of his youth. Those years weren't all bad.

Another folder held scripts to several plays he hoped might be produced. He completed a manuscript titled *Why I Want to Travel,* though it was never published. But his detective novels were the most intriguing, stories like *The Curse of the Gleaming Eye* and *In the Chapel.* He hoped they would be published so he could be an author or someone more artistic than a forensic scientist.

"I am interested in working out those defects in our American life," he wrote his mother, "which breed crime and explaining them to our future parents through fiction."

He enjoyed other unscientific interests, like singing solos for his local social club's choir, studying jiujitsu, and nurturing his large flower garden on his property in the Berkeley Hills. It was all amusing to Kaiser, the reference librarian, who joked that Oscar's passions were eclectic.

"Your special interest in babies, weeds, legal aid societies and spy bibliographies," Kaiser quipped, "shows a cosmopolitan grasp of the world's affairs."

Oscar was ecstatic about his new set of stories. He frequently had a difficult time expressing himself to just about everyone, including his wife, but his frank friendship with Kaiser was the exception, so he sent the librarian the drafts for his opinion.

"I have two, and possibly three novels in pregnant quickening," Oscar wrote.

The scientist was confident about the plots, the structure, and his characters because they were drawn from his own career. "Out of my reading and my experiences there has fruited a guide to expression," he confided to Kaiser. "It appeals to me; it stirs me with desire." His detective stories, now waiting at a printer's workshop, were a cathartic exercise, a venue to tap a section of his brain that he infrequently used in his career.

"I am not wishing to write to win fame or fortune. They are to me nothing more than a guarantee of bread and butter, and my wants for such are well supplied already," he wrote Kaiser. "I want my illusions. I want to play in the fields of fantasy and hunt for the buds of joy. And I want to know whether it be moat or gulf that keeps me from it."

Writing was his joy, and being published would be an honor, so Oscar submitted *The Black Kit Bag* to a major book publisher in New York and hoped for a favorable reply.

Carfare on Thursday had cost him twelve cents. He spent five cents for books that same day and then fifteen cents for bread and cakes on Friday. Oscar's domestic spending logs had been piling up in his lab for years—loads of large, heavy journals each the size of a large bread box, containing thousands of handwritten charts with notes in black ink: "meat—8 cents on Saturday."

Oscar had begun reporting his daily work and domestic expenses in 1910, almost fifteen years earlier, when he first opened his lab in Tacoma. His habit evolved after college graduation from simple bookkeeping to frenetic data gathering. It was all related to his father's death, though Oscar would never admit it. Logs filled boxes in his office, and while it always seemed excessive, his detailed charting became unprecedented by 1923. His collecting tendencies were almost maniacal as the large personal financial logs became more specific and exacting. He logged transactions with clients, especially those who delayed paying him. He sent one client a letter virtually every month for several years, each note polite but stern. As Oscar's business expanded, his routine of gathering immense amounts of data—all different types—became obsessive.

Oscar logged the date, time, and size of each photograph that he took for a case and then created a narrative that explained the facts, described the clues, and revealed his theories. He stored evidence from many investigations, clues hidden in boxes: copper wire from a bomb case, bullet slugs from a murder, and the wax mold he made of a bullet hole from inside a victim's heart.

Oscar paid a newspaper clipping service to ship him a copy of any American newspaper where his name was mentioned. He preserved thousands of periodicals still bound in twine, newspapers that were never read. He kept folders of news stories from each rival expert, with an especially large folder dedicated to Chauncey McGovern that spanned almost fifteen years. There were dozens of articles that mentioned McGovern's cases, ones that didn't even involve Oscar. It seemed to be the criminalist's way of keeping tabs on his chief competitor's wins and losses.

Oscar demanded that his various lab assistants log their hourly work activities, and for almost five years he kept detailed financial records of his mother's expenses in Eureka, Washington. In some of his letters to her, he gently insisted that she send him her bank and tax statements. When his mother began renovating her house, Oscar told her how to have her walls painted and then requested receipts for the work done.

"The paint may be applied to dry walls," he wrote. "This permission will be good for anything costing up to $100."

The financial data had become especially important, because for years he had sent his mother $35 a month. But his information hoarding was not confined to financials or newspapers—he meticulously logged Christmas cards and gifts to and from the Heinrich family. There were graphs on ovulation schedules and notations on every call he made along with the times. Oscar even charted his own urine levels twice a day for one year. He hoped to someday open a clinical laboratory that would handle urinalysis tests and blood work, like the work he did as a pharmacist in Tacoma. All this data collection required incredible amounts of thought and effort, a nearly impossible feat for a man consumed by dozens of complicated cases each month.

Oscar was also traveling more, constantly bemoaning how he missed significant family events like celebratory dinners, summer vacation trips—he was in Oregon during Mortimer's ninth birthday.

"Of course we are glad you are busy but awfully sorry you cannot be with us for the next ten days," Marion wrote Oscar during a family vacation. "Mortimer got the weeps yesterday and insisted he was going home if you did not come up."

Though he had earned $50 a day from his work on the Siskiyou train heist, his domestic expenses were incredibly high. "I hope now that conditions will adjust themselves so that I may take life a little more leisurely," Oscar wrote his mother. "It doesn't seem very long, but it is in fact over a quarter of a century that I have been struggling along under high pressure."

There were heavy expenses to worry about, debts that had forced him to forgo a trip to Washington State to see his elderly mother the year before. "My work has been very light and expense very heavy," he wrote his mother, "and there does not seem to be much left over to go through the Xmas period with."

When a collector wrote Oscar just a few weeks before the Siskiyou train heist concerning an outstanding debt of almost $400, Oscar initially argued over the bill, but then he requested that the lender reduce the debt by forgoing the compounded interest.

"His determination to stick to the face value of his notes and all of the accrued interest thereon is something more than the situation either justified or warrants," argued Oscar.

The strain of being E. O. Heinrich wouldn't relent anytime soon, because the special agents assigned to the Siskiyou train robbery had made some significant discoveries: the killers had secured two secret locations, which contained more clues.

Oscar Heinrich stood inside the small building in the woods, a dingy shack that could scarcely be called a cabin. About a quarter of a mile from the tunnel, the building, nicknamed the Mt. Crest Cabin, had been abandoned years earlier, but it still managed to hang on to the cheap shingles atop its high-pitched roof. A bucket hung from one of the wooden beams while dirty clothes were tacked to the walls.

Railway Special Agent Dan O'Connell seemed a bit out of place in his three-piece black suit, tromping around one campsite, now dubbed "Camp No. 2," about one mile southeast of the tunnel. Oscar snapped photos of charred cooking utensils, a broken wooden chest, and burned beer bottles blackened by a small fire pit. He carried each of the items back to his lab.

The forensic scientist catalogued a towel from the cabin, two suitcases full of clothing with Oregon tags, a burned coat, a valise, fifteen .45-caliber pistol cartridges, a pair of leather gloves, a cap like the one found lying in the tunnel, and socks. There were no obvious clues to the owner's identity, but agents had finally caught a break—they had a name, thanks to Oscar.

The U.S. Postal Service paper that the criminalist discovered from inside the pencil pocket of the overalls was a registered mail receipt for $50 sent by someone in Eugene a month earlier. Agents spelled out the words carefully because the last name was tricky: DeAutremont. Roy had mailed the letter to his brother Verne in New Mexico, and agents quickly learned that there were three DeAutremont brothers in the area during the train heist: Roy, Ray, and Hugh.

The secret serial number Oscar discovered inside the Colt pistol was traced to Hauser Bros. gun store in Albany, Oregon, where a twenty-one-year-old man who called himself William Elliott had purchased the weapon. The dealer's record of sale for the state of Oregon was fairly detailed because the receipt required the buyer to provide his occupation, his address, even his height and his eye and hair color. Oscar suspected that William Elliott was an alias, so he requested writing samples from all members of the DeAutremont family. He squinted at his magnifying glass and determined that the signature matched the example from one of Ray DeAutremont's letters. And despite the problematic nature of handwriting analysis, Oscar was right.

Special agents visited Paul DeAutremont in Eugene, who believed that his three sons were on a camping trip. They described Oscar's physical description of the suspect based on his assumptions from the overalls. It was Roy, Paul DeAutremont admitted—a sad disclosure for a father who was worried about his sons. He handed agents several items, including a red sweater once worn by Roy.

Oscar compared one hair on the towel in the cabin to a hair from Roy's overalls and another from his red sweater—they were very similar, virtually a perfect match. He continued to compare hairs, fibers, and writing samples to create a case against the DeAutremont brothers. The reward had now increased from $2,500 to more than $15,000. Agents added photos and names to the posters.

The manhunt was astonishing: more than two million leaflets were printed in English, Spanish, French, Portuguese, German, and Dutch and distributed internationally. Every post office in America pinned up the posters, while planes dropped leaflets into rural areas. Government agents sent to dentists the descriptions of the fillings in the brothers' mouths, while optometrists were given details about their glasses. Newspapers and radio reports were dispatched daily. Oscar was dubbed by the media "America's Sherlock Holmes," a moniker that privately pleased him.

The hunt went on for months. But soon the media attention waned and the story of the murders of four men at Tunnel 13 near Siskiyou faded from newspapers. By late November 1923, Hugh, Roy, and Ray DeAutremont were in the wind, nowhere to be found and destined to become infamous. Despite the brothers' escape, Oscar's impressive revelations sparked letters from around the world. Professors marveled at his discovery and queried about his methods.

"There is no trick at all to visualize a carload of salt from a milligram of it in a crucible," he wrote a chemistry professor at a university in Lynchburg, Virginia. "It is equally easy to visualize a carload of salt from a single tiny crystal of it under the microscope."

His business became more robust, which was uplifting news for a family in constant need of money, but the Siskiyou train robbery mystery had exhausted him. The forensic scientist moaned about his age, and while he smiled when the media praised him, the escalation of public attention still shook him. He had become distressed after Fatty Arbuckle's case, and over the summer of 1924, he began to lament his choice of profession.

"I feel a strong desire to sit on the bleachers and each other play the game instead of mixing it so hard myself," he sadly told his mother.

He began hinting at retirement, a surprising disclosure from a man who was just forty-three years old. He hoped to pursue other passions, to move away from the horrible crimes he seemed to pursue. Oscar jotted down more ideas for novels as he waited for an answer from the publisher on his short story. And in the fall of 1924, one year after the train robbery, Oscar received a short reply from the editor who read *The Black Kit Bag*: "About this story: it is cast in the usual form of a detective story, but lacks the usual elements of complication and denouement," he wrote. "To my mind the best detective stories are like personally conducted chess games, full of mysteries, surprises, beautiful maidens and heroic youths."

It was dreadful news—and ironic. The brilliant criminalist who had solved some of the most perplexing crimes in America couldn't seem to write a convincing detective story. The editor recommended a college class on short story writing to hone his prose, but Oscar didn't seem interested. Penning fiction was one of the few career ambitions he had failed to meet, so a career as a forensic scientist would have to suffice for now.

"Real Sherlock Holmes, with Four Slender Clews, Pulled Net Around DeAutremonts," read the *News-Herald* headline. It was a flattering article—a tribute to Oscar Heinrich's dogged investigation into the failed Siskiyou train robbery. And now four years later, he was finally receiving retribution.

"The last person in the world one would have picked as a relentless man hunter," said the copy, "stands revealed today as the super-detective whose work brought about the arrest of the three D'Autremont brothers, who bombed a mail train and killed four men in the Oregon mountains in the fall of 1923."

Roy, Ray, and Hugh DeAutremont were caught in 1927 after four years on the run and an intensive investigation that had cost American taxpayers about half a million dollars. Hugh DeAutremont, the youngest brother, had escaped to the Philippines by joining the United States Army. A fellow soldier on leave spotted the youngest brother on a wanted poster in California, and Hugh was arrested in February. Forever loyal, he denied knowing where Roy and Ray were hiding. As Hugh faced a first-degree murder trial in June, the FBI arrested his brothers in Ohio, who were living as Elmer and Clarence Goodwin. The twins had built their lives working in coal mines and mills, toiling at various odd jobs. Ray had married a sixteen-year-old girl, and they had an infant son.

Once Roy and Ray learned about the evidence gathered by Oscar

Heinrich, they both pleaded guilty to first-degree murder. The three DeAutremont brothers were given life sentences, and decades later, they were all paroled. Ray was released in 1961 and lived in Eugene, Oregon, for another twenty-three years until he died at age eighty-four. Hugh left prison in 1958 and became a printer, but he died of stomach cancer soon after.

Roy DeAutremont had the saddest fate, because, more than twenty years after his conviction, prison doctors diagnosed him with schizophrenia. He was transferred to the Oregon State Hospital, the same facility where he had once worked. Surgeons performed a frontal lobotomy to help with his mental illness, but the dangerous procedure left him unable to care for himself, and he died in a nursing home in Salem at age eighty-three.

In 1923, the botched train robbery in Oregon had turned Edward Oscar Heinrich into a legend both in newspapers and within the competitive world of forensic science. He was now the most sought-after criminalist in the country. But Oscar could not help but wonder what else might be out there for him besides, of course, another murder.

8.

Bad Chemistry: The Case of the Calculating Chemist

Holmes was seated at his side-table clad in his dressing-gown, and working hard over a chemical investigation. A large curved retort was boiling furiously in the bluish flame of a Bunsen burner, and the distilled drops were condensing into a two litre measure.

—Arthur Conan Doyle,
The Naval Treaty, 1893

He may have seemed lonesome as he labored by himself inside his laboratory on a hill in a sleepy Northern California town. He slowly stirred the mixtures as the fumes fogged his plastic goggles. His closest friends and colleagues respected him as a gifted chemist, a scientist who enjoyed refining his formulas late into the night as his wife and young sons slept. Rarely did he allow people to peer over his shoulder as he quietly measured chemicals and sparked burners, so no one really understood his struggle with science. His secretary occasionally handed him paperwork for lab supplies, while a night watchman spent evenings patrolling the building's perimeter. But for much of the time he was left with only his solvents and reagents—a determined chemist who vowed to shape history, but in secret. He settled in for a long night's worth of work in July 1925.

A few weeks earlier, Edward Oscar Heinrich picked up a copy of the *Oakland Tribune* and glanced over a small article, just thirty-one words long: "The Pacific Cellulose company, manufacturers of imitation silk, has completed a $50,000 plant at Walnut Creek, and will start manufacturing within a short time, according to Dr. [C.] Schwartz, general manager."

Oscar read it several times. He was intrigued by the idea of such a clever scheme; a formula to create "imitation silk" was sure to become a triumph for the chemist, Charles Schwartz, who had concocted it. In the 1920s, there was a race among scientists to synthesize rubber, timber, and other natural products for commercial use—and none had more monetary value than silk.

Natural silk was pricey because the process to produce it was dependent upon thousands of silkworm caterpillars entering their pupation period; in about forty-eight hours, the three-inch-long pupae would each build a cocoon made of one thread of white or yellow silk averaging about one thousand yards long. Silkworm owners would then preserve that valuable filament by killing the silkworm with hot air or steam, which allowed the cocoon to open. It could take thousands of cocoons to make one woman's silk dress, which would retail for about $25 in the 1920s.

Silk was also becoming an asset in the emerging automobile and aeronautics industries. The first synthetic silks were produced in 1884 from cellulose fiber, but they were too flammable. No one had succeeded in creating artificial silk for commercial use except, Oscar noted, Dr. Schwartz of Walnut Creek. But that name also made Oscar grimace. Just a week earlier, Schwartz had been featured in a very different type of article from the same newspaper—a disgraceful tale of infidelity and extortion. The thirty-eight-year-old chemist was being sued for $75,000 by a twenty-two-year-old beauty from Switzerland, simply because he had broken her heart.

Elizabeth Adam had arrived in Oakland, California, two months earlier and quickly met a handsome, wealthy scientist who introduced himself as "Mr. Stein." They fell in love, and he quickly proposed. The couple made the announcement to their friends, declaring that their wedding date would be set for June 6. But Stein was a cheat, as Ms. Adam soon discovered. His real name was Charles Schwartz, a married father of three boys. Melancholy settled over the young woman, a sorrow that quickly turned to rage. Adam sued Schwartz for "breach of promise," better known in the 1920s as a "heart balm" lawsuit, a legal punishment wielded mostly by jilted women.

For hundreds of years, a marriage engagement was seen as not just a promise but a legally binding contract. When one person broke that agreement (usually the man), the offended party had the right to sue for damages. Men decried heart balm suits as a weapon for scorned lovers. But some fiancées needed financial protection—engaged women sometimes chose to have premarital sex with their future husbands, a taboo during an era when virginity before marriage was coveted. Being left without virgin status after a broken engagement could hinder a woman's chance to find another husband, and a heart balm suit would at least ensure that she was financially compensated. Oscar squirmed as he read that phrase and immediately thought of Berkeley police chief August Vollmer.

Vollmer was embroiled in his own broken heart lawsuit, a scandal that shocked Oscar Heinrich. A woman was suing the country's "father of modern policing" for $50,000, inflicting a very public indignity. She accused the forty-nine-year-old police chief of proposing to her and then marrying someone else the year before. It was an unseemly public scandal, it seemed to Oscar, one that played out in the newspapers and nearly devastated Vollmer's policing career.

"Wife Stands by Vollmer," read one headline. "Declares Cave-Man Tactics Described by Mrs. Lex Not Chief's Style of Making Love."

It was so humiliating for August Vollmer. He had enjoyed the positive media attention he received after creating the country's first police lab in Los Angeles, a project inspired by Oscar Heinrich's incredible feats of forensics during the Siskiyou train heist two years earlier. But this type of publicity was mortifying, and Vollmer declared it a persecution. Reporters even challenged Vollmer to submit to a polygraph exam, the same test he used on William Hightower during Father Heslin's murder case. Vollmer balked.

"The case is far too serious to use spectacular means to keep it alive," Vollmer told reporters. "I have nothing to fear either off or on the lie detector."

Oscar Heinrich was incredulous about the whole thing and mortified for one of his closest friends—a man he revered for his analytical skills and stellar morals. "Please reserve judgment on this for some time," Oscar told his mother. "He is still somewhat unsettled because of the shock and the surprise but I think that he will be able to show a perfectly clean slate."

During his own heart balm suit, chemist Charles Schwartz denied wooing the young woman and then accused his competitors of conspiring to steal his synthetic silk formula. "This is merely a plot to discredit me in my business," Schwartz told reporters. "I defy them to go any further with the action."

Schwartz said that, in fact, someone else had tried to extort $10,000 from him earlier in the week, and he was now fearful.

That could mean trouble for Schwartz, Oscar suspected. A woman scorned might be willing to do anything to punish a heartless lover. But the Swiss woman's allegations were difficult to believe because Schwartz had enjoyed such an impressive reputation. He had been born Leon Henry Schwartzhof in Colmar, France, in 1887, a promising student who earned a PhD in chemistry at Heidelberg University in Germany, served with the Red Cross in Algiers, and became a captain in the French army during World War I. He was wounded during battle and soon met and married a young war widow named Alice Orchard Warden in Derby, England. After the couple had their three boys, Alice Schwartz prayed for financial security; her husband, whom she called Henry, assured her that his invention would make them wealthy.

"Someday, my dear," Schwartz told her, "we will have a great deal of money. Just now you must be patient. My plans are fast coming to a head."

Schwartz was dapper, tall and slender with short, cropped brown hair and an easy smile. He offered charm and wit to his business investors and influential friends, while assuring them that his synthetic silk, created from wood fiber, was a multimillion-dollar idea. His pitch convinced wealthy influencers, along with banks, to invest in the Pacific Cellulose company. He quickly bought an abandoned glove factory in Walnut Creek and converted it to a private office and laboratory.

In July 1925, Charles Schwartz was still troubled over that heart balm lawsuit—Adam was suing him for $75,000 in damages. But he had so many reasons to be optimistic because, he told the owner of his chemical plant, they would make millions from commercial synthetic silk. He just needed time to perfect the formula.

Draped in a yellow dustcoat, he worked late into the night of July 30 inside the second floor of his lab. The gasoline-fueled Welsbach lantern glowed a white light from another room—he had never bothered to equip the two-story building with electricity or gas. The night watchman set out on his scheduled patrol around the property, tasked with catching Schwartz's many competitors, who plotted to steal his secret formula. Schwartz called the watchman to the lab and suggested that he stay far away from the building.

"I plan to do some experimenting with ether," Schwartz warned Walter Gonzales, "and I don't think it would be healthy for you to be sleeping here. Ether, you know, is tricky stuff."

Instead of enjoying a family meal at home, the chemist shared an early dinner with a friend. He complained of a horrible nightmare about a dark man sneaking into his lab and beating him on the head. The visions tormented Schwartz—he was afraid they were an omen. When he returned to the lab, he called his wife and told her to expect him home soon. He phoned his business partner to say he would be finalizing the formula that night—alert the media.

Clear liquid sloshed from a vat onto the floor's wooden planks; Schwartz knew it was volatile, but he ignored the risks. His gasoline lamp flickered and crackled, its light bouncing against one of dozens of windows. Ten minutes later, the walls of the factory shook. The blast ripped the lab's door off its top hinges and shattered its windows. The night watchman raced toward the factory, screaming, "Doctor!" He grabbed a fire extinguisher and quickly broke through the door. Soon, the sirens of fire trucks howled. Flames shot upward from the floor, some three feet tall. The guard gagged from a noxious yellow gas.

Gonzales had responded quickly, but it was too late. Charles Schwartz—the chemist with so many enemies—was now a charred corpse lying on the floor of his own laboratory.

"There was music from my neighbor's house through the summer nights. In his blue gardens men and girls came and went like moths among the whisperings and the champagne and the stars."

In 1925, author F. Scott Fitzgerald's magnum opus, *The Great Gatsby,* was published and immediately became a flop, a victim of poor reviews and terrible first-year sales. A beautifully crafted novel of critical social history, *Gatsby* eventually came to represent the decadence and excesses of the Roaring Twenties.

Across the US, young people were learning a new dance called the Charleston, and the Harlem Renaissance in New York City signaled the beginning of a social and artistic explosion for the African American community, while Art Deco marked the era's modern architectural style.

But 1925 wasn't an optimistic time for all Americans. Farmers still struggled with poverty from the collapse of agricultural prices from five years earlier. Lenders seized farms, leaving rural families destitute. Many coal miners were unemployed, and a record number of Americans abandoned the countryside and moved to big cities as urban areas expanded. And in those big cities, more women were employed, many in clerical jobs.

There was large-scale development of telephones, radios, movies with sound, and more-modern automobiles. Aviator Charles Lindbergh embarked on the world's first solo nonstop transatlantic flight, while improved technology connected Americans across the country. Traditional mores continued to wane, though Prohibition was still the law. There was now an insatiable need for more innovative goods.

During the Roaring Twenties, department stores expanded in big cities, offering customers generous lines of credit or installment plans. Consumer debt was at its highest in history. Inventors like Charles Schwartz found easily accessible loans—their ideas didn't even need to make much sense because there seemed to be plenty of money available.

Before his death Schwartz had explained to his funders that his formula would produce silk at half the price of other cellulose-spun fabrics. He sold stock in the company, but soon investors demanded a salable product. Schwartz stalled for months, offering excuses and requesting more funds.

"I have faith in my invention," he told investors, "but we cannot start production on a shoestring. We'll need more money. This thing cannot be hurried."

In 1925, America was still bullish, but the catalysts of the era known as the Great Depression were joining and strengthening. And soon Oscar Heinrich and the field of forensic science would feel the effects, just like the rest of the country.

August Heinrich's suicide in 1897 had tormented Oscar and his mother for much of their lives. They both fretted over money, but for different reasons. Albertine worried about paying bills, even after she remarried a few years after her husband's death. Oscar became increasingly morose over his own fluctuating income, because even a successful forensic scientist could become crippled under the weight of countless financial burdens. Oscar confided in his mother for years about his money troubles, mostly to explain why he could offer her only a modest monthly allowance.

"I will send you something extra a little later as my money comes in," wrote Oscar. "I have about $2,000.00 outstanding which should have been in over a month ago. Most of them are government accounts so that I can't hurry them up very much."

Oscar and his mother talked about his travels, his sons, and his angst over criminal cases. When Albertine died at age seventy, Oscar lost a beloved parent and a loving confidant, a mother who understood the cost of her son's past fiscal mistakes because she had paid a penalty for her husband's own blunders.

"The funeral expenses of dead horses have cost me close to fifteen thousand dollars," Oscar wrote his mother before her death. "Creditors have remained rapacious, interest has mounted, penalties have mounted."

He would miss his mother terribly, but Albertine's passing also removed another reminder of his distressing childhood. The criminalist, who had been lauded in newspapers around the world for several years by then, felt a bit restless as he reviewed his work calendar. Even though his caseload had seemed enormous since the Siskiyou train robbery two years earlier, he still needed more clients, more money.

Between 1923 and 1925, he had investigated a bobsled accident in Colorado, a bank robbery in Tacoma, and the murder of a police officer in Boulder. He was mentioned in hundreds of articles—his stack of newspaper clippings spilled from his shelves, with many focusing on the train robbery. The magazine *Redbook* featured him in a four-page spread entitled "Manhunt," while a Washington, D.C., paper touted his methods in "Why Criminals Cannot Avoid Leaving Their Tell-Tale Cards." Oscar traveled more frequently during this time, to the chagrin of his family, but Oscar's professional reputation depended on his readiness to report to a crime scene or inside a courthouse.

"My professional calls require me to travel any time on short notice anywhere from seven miles to seven hundred miles or more," he told a friend. "I think no more of taking a train on a half hour's notice for a thousand miles' journey than I do of taking the street car to go to the theater in the evening."

Oscar also turned his attention to professional writing, a necessity for anyone teaching at a university. He published titles like *Checkmating the Forger in Court, The Expert vs. the Alibi,* and *In Re Questioned Documents,* well-received books used by police departments across the country. *The Cooperation of a Library Staff with the Criminal Investigator* was a nod to John Boynton Kaiser, his irreplaceable reference librarian in Tacoma. He proved to be a better academic writer than a novelist, unfortunately. A career as a respected fiction author was still elusive, but perhaps that might change? His dream—his legacy, he thought—would be to write cautionary tales for boys before they had the chance to become criminals.

"I have an idea that I could write stories for boys that would set the stage for life a little more realistically than it does in most juvenile fiction," he once said. "And I'd cast the criminal in the role not of the hero but of the victim."

Boys would surely not want to be victims, Oscar's theory went. But for now, there were bigger worries then publishing children's books. His bank account always seemed depleted, yet he still insisted on a summer vacation. Marion requested a trip to one of their favorite spots, a peaceful sojourn that might calm Oscar's anxiety, even just for a few days. On the afternoon of Saturday, August 1, 1925, the four Heinrichs loaded up the car and drove seventy miles north, to one of the most beautiful places in Northern California, each hoping for a lovely time.

The room was spacious but very bright, almost too bright. Oscar Heinrich glanced at the guest of honor, noting his bright white teeth and blank expression. He peered through the small window at the other end of the room, but there was nothing but darkness.

It was five o'clock—he should have been able to spot the waning sunlight of dusk bouncing off the ripples of the Russian River, where his boys liked to splash on the shore and chase birds. The family's vacation spot in Sonoma County's Summerhome Park was splendid. It was such a shame that he wasn't there that night.

Oscar squinted at the charred body of Charles Henry Schwartz, the eminent inventor, ambitious chemist, and blatant philanderer. Schwartz smelled dreadful as he lay on his back in the undertaker's room in Walnut Creek. His large teeth were bone white in contrast to the remainder of his blackened body. Much of his muscle tissue and flesh was burned off. He certainly looked like the victim of an accidental fire, thought Oscar.

Oscar measured each of Schwartz's body parts, along with his height. Oscar poked the inside of Schwartz's stomach and collected a small amount of its contents, evidence from his final meal. He jotted down the number of teeth and noticed that an upper right molar had been skillfully removed. He carefully lifted each of Schwartz's fingers and noticed there were no fingerprints. Odd, thought Oscar. The fingerprints should have been still visible, despite the fire. He peered at the body's eye sockets and noted that the eyeballs were missing.

Schwartz lay prone with his mouth open, his neck tilted back. The deputy coroner had bound his hands, while his legs were tied with cloth. His knees were bent from rigor mortis. There were specks of black material littered around the table, flecks of skin. Oscar documented the evidence with his camera, producing negatives of three profile shots of the head and two shots of the entire body. His camera focused on the soles of the feet.

"Exposures were made in each case with filter A on panchromatic plate G.D. 12 inch lens at stop U.S. 32 8 minutes each," he wrote.

The criminalist peered inside Schwartz's skull through his mouth—empty. He used surgical scissors to carefully remove part of the scalp. As the skin and hair lay on the table, Oscar squinted and bent over Schwartz's head. He reached for measuring tape and glanced over at the undertaker. Part of the skull was cracked along a fissure as if someone had hit him several times with a hammer or a hatchet. Schwartz had not been killed by a botched lab experiment, like police suspected. He had been murdered. This was officially a homicide investigation.

Now it was Oscar's job to identify the killer. Was it a resentful lover, a begrudging competitor, or perhaps a jealous wife? The suspect list could be long, which was never good news to Oscar.

Murder? Schwartz's wife was hysterical after she had identified the chemist's body. She wept over the details of their last conversation.

"He called me on the telephone just a little while ago," Schwartz's wife told investigators. "He said he had almost finished his work and was coming home soon."

It was certainly her husband, she cried, before collapsing on the floor. The family's physician examined the body and nodded. "I've seen the body and it's that of Mr. Schwartz," confirmed Dr. Alfred H. Ruedy.

Charles Schwartz was dead, Oscar was sure. And a murderer would surely escape the noose if the criminalist and his forensic tools didn't work quickly enough.

The house on Oxford Street was silent at ten o'clock as Oscar opened the door to his office on the ground floor. He stepped inside one of his two dark rooms and flipped off the switch. He plunged the glass plates from his camera into a solution and transferred them to a washing bath concocted to remove chemical residue. By midnight, the negatives were hanging on a drying rack, ready for printing in the morning.

Oscar unwrapped the sections of scalp he had removed from the corpse and placed each piece in separate containers before carefully covering them with a 10 percent potassium hydroxide solution for maceration. The chemical would soften the scalp pieces, making them more malleable for experiments, a technique still used by modern forensic scientists. In other murder cases, Oscar used the same solution to remove the tissue from body parts to reveal bones. After sleeping just a few hours that night, Oscar Heinrich and his assistant stepped inside Charles Schwartz's lair the next day, a dead chemist's own private laboratory.

The science of chemistry had seduced Oscar decades earlier, when the teenager shadowed druggists who blended medicines at the pharmacy where he had worked in Tacoma, Washington. Since then, the forensic scientist had honed his skills in a wide array of sciences, but of all disciplines, chemistry held his deepest appreciation—it dazzled him. And he admired any chemist's journey to discover a new drug, a unique test, or a fresh way to uncover a complicated chemical reaction. Dressed in smocks, gloves, and goggles, chemists handled dangerous substances daily—most were either flammable or poisonous. It was a perilous career at times but one that Oscar valued. Truthfully, he might have been content with toiling inside a chemistry lab for the remainder of his life, but forensic science drew him into a more lucrative, intoxicating world.

In early August 1925, as Oscar wandered around Charles Schwartz's massive lab in Walnut Creek, he began to visually reconstruct the scene. His first assignment was to answer the most crucial questions: How had the fire begun and where? A member of the district attorney's staff and the deputy state fire marshal eyeballed glass beakers and crouched near mysterious liquid on the floor. They reviewed their suspect list, which included Schwartz's furious fiancée and the envious chemists in Europe who had threatened him. Suicide was briefly mentioned, but he was beaten on the back of the head. His wife subscribed to the theory that murderous rival scientists had killed him.

"He was murdered by people who wanted to get his secret formula," she told the sheriff. "I'm sure they set fire to the plant to hide their crime."

Oscar wanted to learn more about the chemist—by profiling the victim, he might be able to uncover his killer. The criminalist searched for items to test: beakers, burners, test tubes, and books, but scientific equipment in the laboratory was sparse. Odd, Oscar thought, because those items were necessities to any chemist. And there were no gas or water sources, so how could he work on his formula for synthetic silk? The criminalist glanced around the lab again. He called over the investigators.

A fraud, Oscar declared. Schwartz was duping lenders. There was no recipe for synthetic silk, and there never had been. Schwartz wasn't an enterprising chemist. He was a con artist.

Schwartz's Ponzi scheme was a dicey swindle. It was also an excellent motive for murder. Oscar picked up his pad. His list of suspects had just lengthened, and there was so much more work ahead.

He looked down at the floor's wooden planks and ordered his assistant to start sweeping while he sifted through the debris. Oscar was certain that Schwartz was a highly intelligent grifter, and now he needed to know more about the blaze. Had Schwartz been murdered before or after it was ignited? Where had the killer been standing? Oscar asked the fire marshal, the first investigator on the scene, if there were any flammable materials in the lab on the night of the fire.

"Five gallon can of carbon disulfide," he answered.

Actually, there were *two* cans of the colorless, volatile, and flammable liquid, and there were more cans lined up beneath a bench where Schwartz's body was found. Oscar fiddled with the gasoline lantern on the shelf in an adjoining room and then examined the doors to the lab. One of them was open at the time of the fire, while the other was locked.

"A Sikes office chair was in the path of the combustion," he observed. "Shows charring of the varnish completely covering the chair including the underside of the seat."

He asked himself a series of questions about the origin of the fire, a self-reflective exercise that any detective would use. "Did the explosion start from the gasoline lamp and later spread to the laboratory," Oscar wrote down on a notepad, "or did it start in the laboratory and later reach the lamp?"

He stepped into the hallway, just outside the main doors of the laboratory. "Cracks in flooring on east and west side of door show a running through cracks of a volatile combustible liquid poured on the floor," he noted. "A match dropped on the floor."

The fire was started *outside* the closed doors and had crawled underneath the door—Oscar realized that the killer had hoped to trap his victim. Maybe Schwartz was killed *after* the fire started because the blaze didn't move quickly enough? He still needed more information.

Oscar turned back to the area where the body was found—someone had started a fire there, too, because he discovered rags soaked with benzol. He examined the floor and decided that the fire was at its hottest by the body. "By tracing the flames on the floor, I found that the fire had not started in one place but five different places," he told the sheriff.

This proved that Schwartz had been dead *before* the fire began, Oscar realized. There was a fire directly under the chemist's body. It was a dreadful ending to a sad tale—Schwartz was a lothario, a cad, and a con man, but no one deserved to die so terribly. He had been bashed on the back of the head and then set on fire like a piece of garbage. It was time to conclude this investigation.

Oscar looked around the room and spotted a storage closet. He wondered if Schwartz had been inside the small room before he died. He pulled open the door, then retrieved a spray bottle. The mist lightly landed on the wood, and the luminol droplets glowed on the door. Oscar blinked and adjusted his glasses. Large sprays of blood. "Projection group—consists of several large stains ranging from ¼ to ⅜ inches in diameter projecting forward and downward in an angle of approximately 45 degrees toward the inner end of the closet," he wrote.

There was an *incredible* amount of blood, so much in volume that it had dripped through the closet floor and soaked through to the ceiling of the office below. It was probably a head wound, just like the one Schwartz had received from his killer. The chemist may have stumbled inside the closet before he staggered to the workbenches where he died.

Oscar poked the blood sprays—the red dots flaked. They were long dried—the fire had not been anywhere near the closet. Schwartz's body was likely secluded in the closet for hours before the fire. He scraped off samples of the blood and wrapped them carefully in paper for testing later. He made notes and then paused.

He scanned his notes on the interview with the night watchman.

Something wasn't right—there was a problem with the timeline. He found the detail. The guard told investigators that he had seen Charles Schwartz just *ten minutes* before the fire. If that were true, thought Oscar, then how could the chemist's body have been inside the closet for hours before the fire? The criminalist had an idea, and if he was right, it would horrify Americans across the country who had followed the case of the murdered chemist.

When police asked Schwartz's wife to identify his body, she had recognized his watch, a sure sign that the man on the table was her husband. The night guard agreed that this was his boss, because he had watched Schwartz empty his pockets and count his change that night—and the exact amount of coins were found in the victim's pocket. What else might help identify her husband? asked police. His missing tooth, replied Mrs. Schwartz.

Charles Schwartz's dentist had just removed an upper right molar a few weeks earlier, so Oscar reexamined the body. There was indeed a molar missing. He took a cast of the body's teeth.

"That's the tooth I extracted not very long ago," confirmed Schwartz's dentist.

And then Oscar embarked on a gruesome task. Investigators had recovered the victim's right eyeball inside the laboratory—the sheriff handed it to Oscar.

"Replaced the eye in socket prior to making photographs," he noted.

When Oscar later dissected the eyeball, he noted that the iris appeared to be punctured, perhaps with a screwdriver. But why? Oscar looked at Mrs. Schwartz's interview notes. He found her statement about a bizarre burglary at the family's home during the night of the fire.

"Every photograph of my husband was gone from the house," she told police.

Someone was trying to prevent the police from identifying the corpse, Oscar surmised. Still—the findings seemed to tally with Charles Schwartz's physical details. Or not. Oscar smiled at the report on Schwartz's physical exam sent by a life insurance adjuster.

During most murder investigations, detectives pray for at least one aha moment, a key discovery that cracks the case. Oscar Heinrich was about to present this group of detectives with a chest brimming with key discoveries.

Charles Schwartz was a slight man at just five feet, four inches, according to the report. Oscar reviewed the corpse's measurements—the man lying on the coroner's table was actually three inches taller.

Oscar quickly scanned his notes from the fire marshal, which reported that the body was in rigor mortis when firefighters found it—knees bent. The stiffening of the muscles and joints wouldn't usually set in until at least several hours after death. Again—he remembered that the night watchman had seen Schwartz just ten minutes before the fire. Oscar gathered more data. He turned to Contra Costa County sheriff Richard Veale, who was leading the investigation.

"I'd like to know, first of all, what Schwartz ate for dinner the night of the fire," Oscar asked.

"Do you think that makes any difference at this point?" the sheriff replied. "Then I can tell you that he ate cucumbers and beans."

"There is nothing in the dead man's stomach but some undigested meat," replied Oscar. "And there's not the slightest trace of cucumbers and beans."

The criminalist had received a sample of Schwartz's hair from a brush a few days earlier and compared it to the hair from the scalp of the corpse.

"Under the microscope I found them to be entirely different," Oscar said. "There isn't even the slightest similarity."

Schwartz had a prominent mole on his ear, while the corpse had none. The fingerprints of the victim had been removed with acid, which would have been easily accessible to a chemist. But both Schwartz and the corpse were missing an upper right molar, argued the sheriff. Oscar pulled out a photograph of the tooth cavity.

"The tooth hadn't been extracted at all," he said. "It was knocked out with a chisel. You can see the root still imbedded in the jaw."

The tooth had been removed *after* death. Besides those clues, the body's condition during autopsy made it seem unlikely that Schwartz had died inside a lab filled with fumes from carbon disulfide.

"I asked myself what effects these vapors could have had on the victim," Oscar told the sheriff. "If he had battled for his life against them, a lung hemorrhage would have resulted. The autopsy revealed no such condition. This isn't Charles Schwartz."

The police seemed stunned, and Oscar Heinrich felt satisfied. He had outwitted the calculating chemist.

Later that day, investigators gathered to review Oscar's findings, still trying to sort out how to present the story to the media. Berkeley police captain Clarence Lee, one of August Vollmer's top cops, quietly listened to the evidence. He quickly felt nauseated, and Oscar soon discovered why.

Lee and Charles Schwartz had known each other for years, ever since the chemist became fascinated by the cop's work at the state's Identification Bureau, the agency charged with collecting data on criminals and creating records. For three years, Schwartz enjoyed dropping by the station for casual chats with investigators; he explained that he was a local chemist who fancied himself a student of criminology.

"He tried to give the impression that he worked as a detective in Europe," Lee told Vollmer, "but when I asked him about his experiences, he was always vague in his answers."

Lee and Schwartz chatted about historical crimes, cases in the news, and various investigative techniques. But now Lee grimaced as he recalled one frank conversation in early spring. It had been a debate that began when Schwartz leaned against Lee's desk at police headquarters and mused about murder.

"What interests me about murderers, captain," explained Schwartz, "is their lack of forethought. They simply do not think."

"Sometimes they don't have a chance to think," replied Lee. "Murder isn't as easy as it looks. Very often it's just a single bad break in the luck that ruins the whole scheme."

"I grant you that," Schwartz agreed, "but I am not talking about sudden crimes of passion. I mean those cases where the prospective murderer actually does plan the crime long in advance."

Lee had only half-listened to Schwartz's ponderings much of the time, but months later he recalled one strange musing in particular.

"It's amazing to me the way they fail to cover up their tracks," said Schwartz.

Captain Lee had not been alarmed months earlier by the chemist's blunt comments, but he certainly was now. Investigators began hunting victim-turned-killer Charles Schwartz using Oscar's clues. Catching the murderer was crucial, Oscar agreed. But if he couldn't identify the man lying on the coroner's table, this case would forever plague him.

The forensic scientist was exhausted when he held up the pamphlet to the overhead light in his Berkeley laboratory and stared at the print on the cover: *The Philosophy of Eternal Brotherhood.* It had been found on the body. There were light pencil markings above and below the title. Bloodstains had sprayed on the back cover.

"Two directions of splash shown," Oscar wrote in his work journal, "as if book shifted between two successive projections, or bleeding object shifted."

Schwartz's secretary revealed two more of his secrets to police. She had watched the chemist shove $900 into his pocket just hours before the fire; more crucially, she said that months earlier, Schwartz had purchased an advertisement in a San Francisco newspaper for a laboratory assistant, a likely ruse to lure a stranger to visit his company's building. Charles Schwartz took out several life insurance policies, double indemnity policies covering accidental death that totaled more than $185,000. If he died, his wife would be left with the majority of the money. Oscar declared that this was likely a case of insurance fraud.

He held up another religious pamphlet found on the body: *The Gospel of John the Apostle*. He could see a faint signature at the end of a confessional phrase, markings unreadable to police investigators—but not to a forensic scientist. Oscar sprayed a chemical on the booklet, and the faint pencil markings grew darker. Under a microscope, he could read the signature "G. W. Barbe." He compared the pencil writing from the first booklet, and they seemed to match.

The victim carried with him a small collection of religious literature with numerous passages underlined. He was a traveling preacher, Oscar guessed, a poor man who spread the gospel as he crisscrossed the country. Instead of using his routine approach to profile the criminal, Oscar created a profile of the victim. Lying near the body was a bindle with a small sewing kit and a bar of soap. A bindle was a cloth bag hanging from the end of a stick that could be rolled up after coffee grounds were used—something that a drifter would carry. The victim was wearing worn socks and clothing, but the man's short haircut and well-cared-for hands and feet meant that he was not a typical homeless man.

When newspapers across the country printed the profile of the murder victim, an undertaker in Placerville, California, told police that he had known a preacher who had answered a newspaper ad for a laboratory assistant. Gilbert Warren Barbe was a World War I veteran, a college graduate . . . and a traveling preacher. He had perfectly fit Oscar's victim profile. Meantime, Charles Schwartz was still a dangerous murderer on the run, a killer chemist who hoped to outsmart police by murdering an innocent man and then faking his own death.

As police continued to investigate Charles Schwartz, his public image evolved from scientist to scoundrel. Heidelberg University in Germany denied that he had ever received a chemistry degree, though police did find a certificate that indicated some study at a minor university in France. He had earned just enough education to make his elaborate plans seem plausible to investors.

His wife accused him in the newspapers of being a louse, a serial cheat who had begged for forgiveness when his mistresses confronted her. The woman who sued Schwartz, Elizabeth Adam, said he had gifted her with more than $1,000 during the course of their relationship. She denied that she had been intimate with Schwartz, no doubt to protect her own reputation, but he had bragged about their liaisons. Schwartz's former employer accused him of stealing a bottling machine along with almost two thousand pounds of scrap iron. Schwartz had also threatened a co-worker with a .25-caliber automatic pistol. The press eagerly reported on it all as the manhunt for Schwartz continued.

With all that attention, he couldn't stay hidden for long. After his "death," Schwartz had rented an apartment in Oakland under an assumed name—Harold Warren—creating the persona of a handsome structural engineer who enjoyed entertaining guests with elaborate meals and card games. And that's how Schwartz spent his final week of freedom, attending parties with neighbors that included loads of laughs and losing at games of cribbage. But on August 3, American newspapers revealed that Schwartz was suspected of murder. Soon, one of "Harold Warrens'" new friends recognized Schwartz from a newspaper photo. And on August 9, Berkeley police captain Clarence Lee, Schwartz's old acquaintance, tracked him down at his secret apartment around two thirty a.m.

"Open that door—police!" Lee yelled.

As the police captain kicked in the back door, he heard a shot. He rushed to the living room and found Schwartz lying on the floor with blood pouring from his right eye. A German pistol lay nearby as he gasped—soon his breathing slowed and then stopped.

A suitcase nearby was partially packed with books and maps inside that hinted he had hoped to travel to Mexico or South America. Poison tablets were beside his bed, and police discovered a tearful note left for his wife. As always, the chemist was determined to dictate the details of his own story, including its finale. In the letter, he told his wife that Gilbert Barbe had indeed visited his lab, hoping for work. When Schwartz turned him away, the preacher suddenly attacked and the chemist beat him to death in self-defense.

"The only thing I did was I tried to burn him, to wipe out and go," wrote Schwartz. "I kiss this spot in bidding and kissing you goodbye. My last kiss is for you, Alice."

It was a tragic, pointless ending for a killer who had once brimmed with such promise and potential. And no one was satisfied. His wife eventually recovered some of the money after several court fights with the insurance companies, but after legal bills, it was unlikely she was left with much.

Schwartz's account of being attacked in his lab was deemed unreliable, but there would be no judicial justice for Gilbert Barbe or his family, either, though he was given a grand military funeral. As the Schwartz case concluded, investigators wondered about possible accomplices, perhaps even his wife, but police captain Clarence Lee shook his head. "Schwartz was too familiar with crime detection through his constant interest in work being done in the identification bureau to dare the risk of accomplices," he replied.

And Oscar Heinrich agreed, based on the elaborate murder plan. "It was too perfect to have been the work of more than one," he concluded.

The brightest killers, Oscar concluded, were solitary criminals.

And thirteen years after Schwartz's death, the DuPont company, a massive American chemical conglomerate, began marketing the first commercially available artificial silk—a material called nylon.

Lee remembered that strange conversation he had had with Schwartz at the Berkeley police station, when the captain had argued that a murderer's sophisticated plans would be no match for foul luck. Schwartz had disagreed.

"The criminal who leaves room for a chance mishap is not planning his crime properly," insisted Schwartz. "He must not hope for the breaks—he must force them himself. It is all, to my mind, merely a matter of proper planning."

But Schwartz never imagined that even proper planning would be no match for a chemist turned criminalist like Oscar Heinrich.

The forensic scientist's remarkable results in the Schwartz case provided a thrilling victory for Oscar—more time in the media's spotlight. "Making the Dead Speak," read one headline. "The Man Who Makes Murder Out," declared another.

But Oscar's life was becoming frustrating. He feared he might never be able to retire, he lamented to his mother before she died, because of "some expensive mistakes [that] had been made in the past." He continued to lie to his wife about their money woes, leaning on the belief that it was "a husband's duty" to provide for the family in the manner to which they'd become accustomed . . . even as that became an ever more impossible goal. He became more resentful with each new case. Criminals sickened him, but he felt obligated to the innocent people who deserved his protection. And he needed money to support his two sons, who would be attending pricey private schools soon. He needed a higher income, perhaps the steady, reliable salary of a full-time chemist. But the draw of crime scenes and solving the unsolvable cases was too strong. Oscar refused to quit forensics, despite every instinct in his bones.

9.

Bits and Pieces: The Case of Bessie Ferguson's Ear

You have come up from the south-west, I see. . . . That clay and chalk mixture which I see upon your toe caps is quite distinctive.

—Arthur Conan Doyle, *The Five Orange Pips,* 1891

The saltwater marshes in El Cerrito near San Francisco were almost mythical, with swamps of black mud so thick they could hobble a workhorse. During sunsets, the dense patches of tule stalks virtually glowed in the warm light. It was August 23, 1925, around six thirty p.m., and less than three weeks earlier, Charles Schwartz had killed himself.

Roger Thomas stepped gingerly toward the shore. His father had tasked the twelve-year-old with collecting some of the beautiful, hard-stemmed plants for decoration inside their house.

The boy walked carefully toward a section of green tule just along the shore of the bay. Amid the grasses, lying atop the muddy ground, was a small light-brown-colored object—perhaps a tiny sparrow dead from disease. Roger stepped toward it and stopped. He yelled for his father, who came running. He stared down before calling the police. The carcass of a tiny bird would have been far less troubling than what was actually lying in that marsh. It was certainly a murder case, and Oscar Heinrich, as usual, would soon be at its center. He vowed to solve this murder—he was sure he could do it. As with every investigation, his reputation depended on it.

"New murder case," read the note that the criminalist jotted in his journal on August 24, 1925. He slipped a white lab coat over his dark suit, settled into his chair, and slid over a microscope. Now he was forced to look at the clue once again, fully examine it—a horrid task, even for a trained investigator. In many murder cases, Oscar found that either the killer or the victim would seem fascinating to him. In this case, it would be both.

The sky had darkened outside as Oscar snapped rubber laboratory gloves over his hands. The Berkeley Hills neighborhood was virtually silent at night after families flicked off their lights, but inside his cavernous lab, Oscar prepared to solve a dreadful case involving an enigmatic victim who would trouble him for the rest of his career.

The criminalist stared down at the brown object that the boy had discovered the day before in the tule swamp of El Cerrito. A dainty, delicate ear slid from his sanitized container and landed on a sheet of white construction paper. He squinted at the matted light brown hair—the ear was attached to part of the scalp. It rested there on his desk like a trinket sent by a macabre admirer. Oscar leaned toward it, adjusting his spectacles.

The ear's exterior was covered with small light freckles, likely from years of sun damage. He noted the ear's size and shape. There was a tiny hole in the lobe for an earring—detectives had missed that detail. So this was likely a woman's ear. Oscar placed her scalp on some paper with a grid and reached for his measuring tape. He scribbled down some numbers and gently lifted sections of her hair—no evidence of blood clotting.

"Cut after death," he wrote. "Dismembered."

But why? Oscar knew that many times a killer dismembered his victim to prevent identification. Maybe he knew her? Searchers found another segment of scalp nearby, wrapped in a month-old newspaper. Investigators were canvassing the neighborhoods near the swamp, hoping to find witnesses.

Detectives entered every duck blind and hunting lodge along the shores of San Francisco Bay and tiptoed along the edge of the swamp where they discovered some women's clothing that had been cut into small pieces and wrapped in Oakland newspapers.

Oscar sensed urgency from investigators because the bloodthirsty media would soon pick up on the story; the violent tale of a depraved killer would surely frighten all of El Cerrito.

In his lab Oscar picked up a hatchet laying on his table, a sharp weapon discovered inside a cottage a half mile from the swamp. The owner claimed she hadn't stayed there for more than two weeks; she said she had been rattled by the crimson spots on the blade.

"I do not know how blood stains got on to it, for I have not killed chickens, rabbits, or anything else," said Iva Graham. "How blood got on to it, as the police say, is a puzzle to me."

Oscar fingered a small bottle filled with clear liquid and "benzidine" written on the side. Using an eye dropper, he dribbled it onto the edge of the hatchet and waited for the red liquid to turn blue, signaling blood. Nothing. Even if the killer *had* hidden inside the cottage, he hadn't used the weapon on his victim.

Oscar worked for nearly twelve hours inside his lab that day, assembling a composite profile of the victim: a woman in her twenties, naturally blond with hints of red and brown, a lady of some refinement because of the high-quality texture of her hair and skin. Oscar could tell that she took pride in her appearance. She was of Scandinavian descent, he suspected. Based on the wound, he estimated that she had died a week earlier.

A knock came on his laboratory's door, and standing in the doorway was Contra Costa County sheriff Richard Veale, who had been the lead investigator on Charles Schwartz's case weeks earlier. Oscar smiled. He and Veale had communicated well on that case, and now they would be working together again. August Vollmer also sent Captain Clarence Lee, the cop who had chatted with Charles Schwartz as he secretly pondered how to commit the perfect murder. Oscar frequently worked with the same investigators, especially in cities much smaller than Los Angeles or San Francisco. Berkeley boasted a strong cache of detectives, thanks to Vollmer, but there were just a few who could be trusted with major crimes. These were two of his favorite cops. Oscar wondered to himself: Who was this woman, and where was the rest of her body?

With metal tweezers, the forensic scientist retrieved several fly larvae, which he found developing in the fatty tissues underneath her ear. These tiny insects would help him estimate her time of death. He flipped open *The Control of House Flies by the Maggot Track,* a short book he had received seven years earlier from John Boynton Kaiser. The reference librarian delighted in new scientific techniques and realized that the book might help Oscar determine the time of death using bugs.

The study of insects in crime solving, known as forensic entomology, had been used since the mid-1800s by a handful of scientists in Europe who hoped to use nature as a co-investigator, but the method had never been documented in a criminal investigation in America—this case would be the first.

There are two primary ways to determine time of death, or postmortem index (PMI), using insects. The first is to use the secessional wave of bugs. In Kaiser's book, Oscar read that *Calliphoridae* (known as blowflies) typically are the earliest of all insects to respond to a decomposing body; much of the time they lay their eggs on the corpse within twenty-four hours. Other species, like beetles, arrive only after advanced decomposition. Oscar found only blowflies on her ear, signaling that she had been recently killed. He was relieved—the longer a body was missing, the harder it would be to find . . . and Oscar desperately wanted to find the rest of this woman.

He also used a second method in forensic entomology, which was to measure the larvae's age and development, their relative age. These insects were in their earliest stage, so the murder was recent, perhaps within forty-eight to seventy-two hours before he had received the ear in his lab. It was a remarkable claim that few criminalists at the time could have made. And then Oscar entered the mind of the killer, the only way he could discover who would have dismembered a body and then hid its pieces.

"Assuming an additional twenty-four hours to be required to prepare the body and clothing fragments for distribution," he wrote, "I estimate the time of death as approximately ninety-six hours before the afternoon of August 24th."

Oscar had to narrow down that timeline even more. If someone had *just* killed someone and wanted to dismember her body, he thought, it could take quite a long time. When might he have the most time to work undetected? The darkest night of the week. Oscar flipped through a local newspaper until he found a table of tide and moon phases over the past few days.

"The murder and distribution could be performed unobserved at any time after 8:30pm of August 19th, 20, 21, and 22," Oscar wrote.

Within twelve hours of being handed just one dreadful clue from the swamp, Oscar had created a profile of the victim and then estimated when she had been murdered. Now police tasked him with directing them to the rest of her body. He picked up the clothing found in the neighborhood near the swamp, a brown jacket with a fur collar that likely belonged to the woman. Using a magnifying glass, he discovered wet sand throughout the fabric, a key clue. He turned to a reliable tool, one of his most favorite. He slid his chair over to the petrographic microscope, the same one he had used in Father Heslin's case four years earlier. "Small fragments of plaster, coal, decayed redwood and similar debris which leads me to believe that burial of the clothing during temporary concealment of it was under a house," Oscar noted.

He was coming closer to finding her. The sand was his best clue, evidence that might lead police to her body. And now the press was adding to his strain.

"Tule Swamp Drained to Find Body," one headline read. When

reporters realized that "America's Sherlock Holmes" was working the case, they hounded him for details—he ignored them all. Oscar was still sorting out his relationship with the press, which was shaky and even hostile after years of inaccurate reporting and unflattering profiles.

"The city editor is, without exception, a spud-bug and a road-hog on news," he complained to John Boynton Kaiser about one newspaperman, "and his reporters scurry around like a nest of road lice getting it in . . . ready on the instant to apply the powerful screws of the great press they represent, like cockroaches."

Oscar cooperated with a select number of reporters, and he often praised papers like the *San Francisco Chronicle* for being judicious with their crime coverage. But he was still suspicious of the media even when he was at the center of a case with a favorable outcome, like the Siskiyou train robbery, and he viewed the press as a tool, not an ally.

"Because of the higher presence under which they work, their chestiness over their jobs, their intense rivalries," he wrote Kaiser, "their suspiciousness, their cynicism, they require careful handling."

Ironically, Oscar didn't seem to realize that he was facing those same challenges with his own professional competitors, the forensic experts who were his antagonists during dramas that would be on display in American courtrooms for the next two decades.

Oscar revealed bits and pieces of new evidence, and investigators snatched each one to feed to the press. The sand from the victim's clothing contained small particles of clamshells. There were enough shells to suspect that the sand originated *near* the ocean, but not enough to have come from a seaside beach.

Oscar stared at the grains. What type of water was near an ocean in this part of Northern California? There was no sand hidden within the tule stalks of El Cerrito, just black muck. He pushed his microscope to the side and grinned.

"Size of rock particles indicate alternate current and periods of quiet with deposit of water borne material," Oscar wrote. "Suggest backwaters of stream emptying into tidal waters."

The sand did not come from the nearby beach in El Cerrito. It came from water that flowed into the ocean; Oscar was certain. He dug into a drawer and pulled out a United States Geological Survey map, a document that would show the regions where that type of sand might be found. He studied it, shifting his eyes back and forth between one particular location and El Cerrito. It seemed too unbelievable to be true. He circled one area with a pencil and then braced for dozens of questions from hesitant detectives.

"It's somewhere around Bay Farm Island," Oscar told them.

They stared back.

"That's fully twelve miles from El Cerrito, where we found all this," one of the detectives argued. "It's even in another county."

Oscar's explanation was slow, thorough, and confounding to the investigators. He described the results from his petrographic test—the woman's clothes contained just a tiny amount of salt and chloride (like ocean sand) but quite a lot of freshwater vegetable matter and chemicals—like sand from a bog. The detectives seemed to follow his logic.

"From the size of these grains of sand I determined that they came from a spot where there were fresh-water gullies and where the movement of the water over the land was slow," he explained.

The sand came from slow-moving water, and that's what he wanted to find on his geological survey map—that spot.

"The most likely place was Bay Farm Island," Oscar explained. "The island is separated by Alameda by a slough and into this the fresh water of San Leandro Creek flows. That accounts for the fresh water element I've talked about and would provide conditions to produce sand similar to that which I found."

Oscar didn't know it yet, but his methods would make forensics history once again. He had used yet-unpublished methods of discovering quartz grain surface textures to point police toward a new location. It's a technique used now by modern forensic geologists, but Oscar didn't have the benefit of using today's atomic force microscopes.

Those devices create high-resolution 3-D images using a small probe with a sharp tip that scans back and forth across a grain of sand to measure the surface topography at up to atomic resolution. Instead of using a spacial database on a computer to chart the body's coordinates, as current researchers might, Oscar used his geological map and plotted out the origin of the sand—a remarkable technique. The detectives looked at each other. Oscar Heinrich was rarely wrong, they concluded. The men prepared to issue new orders to the searchers.

If El Cerrito was a lush marsh, Bay Island Farm was a muddy bog. Police and volunteers slipped and slid along the mudflats under the drawbridge, hauling shovels and spades with the intention of unearthing a corpse.

Newspapers printed Oscar's description in their headlines: "Victim Young, and Woman of Refinement, Experts Declare after an Inspection of Clues." Police desperately searched for the identity of the young, blond, well-kept woman of Scandinavian decent. And in less than twenty-four hours, they would find her.

The forensic scientist had spent the last few years admiring his sons as they studied in the rooms above his lab on Oxford Street. One boy could be found with his nose in a book, while the other seemed to prefer daydreaming. Oscar's assessment of their future would shape their lives, to a certain extent. One son would feel immense pressure to equal his father's success, while the other would struggle throughout his youth to satisfy his father's high expectations. Theodore, who was fifteen years old in 1925, mimicked Oscar in most ways.

"He is well balanced, quiet, courteous, and helpful," Oscar proudly told his mother before she died. "In every respect he is a very agreeable boy to have around. For a young boy he has an unusual range of expression and quite an ability to write."

But eleven-year-old Mortimer served as a sort of friendly foil to his older brother.

"Mortimer persists in being the opposite of Theodore," wrote Oscar. "He is quick at learning, but it is somewhat hard to keep him at it."

The complicated relationship between Oscar and Theodore, as they both aged, would impact the criminalist's life. But he adored his eldest son.

"My dear big boy," Oscar wrote Theo on his fifteenth birthday. "Your birthday brings to me pride in your growth to new duties and responsibilities and fond recollections of your very earliest days with your mother and I."

Oscar's interests away from his lab continued to multiply. Despite a blunt rejection from one fiction editor in New York, he had not abandoned the idea of publishing a detective story. He moved on to another publisher, vowing to never give up—his mantra in life.

"The book which you remember and concerning which you have so kindly inquired is not yet written," Oscar wrote a friend. "There never has been any difficulty about a publisher for it. In fact, Appleton is waiting for something from me now."

Crafting stories to both titillate and educate readers would motivate him to carry on at his writing desk. He seemed to hope to die with a pen in his hand, not a magnifying glass.

"As you know," Oscar wrote to his friend, "writing maketh an exact man."

It would be decades before he would learn whether that fantasy would become real.

Her tibia lay next to her fibula. A measuring tape stretched out alongside her humerus, her upper arm. It was easy to lose count of so many tiny bones. She had been scattered across the mudflats of Bay Farm Island, precisely where Oscar Heinrich predicted she would be found, about twelve miles from El Cerrito.

Two searchers carrying shovels found the skull pieces buried beneath a drawbridge by a river. Small clamshells adhered to her bones. When Oscar received the evidence at his lab, he felt satisfied, even relieved. Once again, his unique methods worked, separating him from the lesser investigators who claimed to be his peers. No one else could have accomplished this, he crowed to himself. Oscar removed his tweed jacket, rolled up the sleeves of his white dress shirt, and snapped on his dark rubber gloves once again.

"Examined a skull which has been cut in several fragments with a saw and cast into the waters of San Francisco Bay from the Bridge leading to Bay Farm Island," Oscar wrote. "Killed by a blow on the top of the head with a blunt instrument and thereafter her body was dismembered and cut up into small units and these units distributed over Alameda and Contra Costa County."

Oscar slid her upper and lower jawbones in place, completing the skull. Her head faced toward the ceiling with her jaws wide open, as if she was killed in mid-scream. A pair of schoolboys unearthed a bag containing her kneecaps, ribs, and other bones, meaning that someone had tried to remove her flesh using chemicals. The killer also buried a large piece of her torso under the bridge on Bay Farm Island, along with a lung. Police found a piece of a woman's breast and her abdomen in Rodeo, thirty miles north. The El Cerrito marsh where the scalp was found was about halfway between the two locations.

"It may be tentatively assumed that dismemberment was effected near the scene of the attack," Oscar noted.

The murderer likely killed her and then mutilated her body in the same location to avoid being caught with a corpse, he thought. As Oscar finished reassembling the mystery woman, an Oakland dentist confirmed her identity using a customized porcelain crown in her lower jawbone: Bessie Ferguson. She had recently been reported missing.

Oscar stood to the side while her mother wept. After her family left, Oscar stepped back and surveyed his ghastly jigsaw puzzle.

He had assembled a nearly complete skeleton of a woman who was once beautiful, provocative . . . and devious. Oscar had never met someone quite like Bessie Ferguson. But now he depended on her to reveal her killer.

Her parents were bereaved but perhaps not that surprised—they had been concerned about their thirty-year-old daughter for years. The Fergusons were a devoted family with four girls (including one who had died the year before) and three boys. This was a nightmare for her parents.

William and Annie Ferguson shook Oscar's hand inside his Berkeley laboratory. He asked about their family background, searching for insight . . . and suspects. Family squabbles that turned deadly were not all that unusual. And the Fergusons' history was fraught with tragedy.

Bessie's father was a flour miller until he had been paralyzed after an accident five years earlier. Her older brother, Will, was a thirty-seven-year-old apprentice miller who suffered a stroke after separating from his wife. He struggled as a single father, so Bessie lovingly cared for him.

She was educated, studious, and very attractive, with a styled bob haircut and curvy figure. But somewhere along the way, Bessie's plan for a wonderful life had shifted. It might have started to veer off course at age twenty-seven, when she married Sidney d'Asquith, a gambler who loved to bet at the racetrack.

"Kept the family broke," wrote Oscar after talking with her mother. "Lived together for 2 or 3 years."

Bessie was trapped in an unhappy marriage when she began working for a handsome San Francisco businessman, an employer who would later become her lover. She eventually became a nurse, but she hadn't worked steadily for at least three years. Bessie was living alone in

Oakland, the victim of a string of bad choices despite a supportive family. And soon she would make one final, grievous mistake.

Despite being chronically underemployed, Bessie Ferguson always had spending money, her mother noticed. She was naturally modish, hauling around expensive bags, stylish clothes, and even a diamond ring.

Her mother suspected she was playing a tricky game by carrying on affairs with several men. She used various names, including "d'Asquith" and "Mrs. J. Loren." One paramour was an accountant, her former employer in San Francisco. There was also a dentist and a physician, both in Oakland. She stashed love letters and provocative notes inside a trunk, in case she needed leverage on her lovers later on.

Many people had love affairs, of course, without becoming victims of murder. But most people weren't quite as calculating as Bessie Ferguson. For more than seven years, she extorted money from at least three men, each of whom was paying for a child who didn't exist. One "father" suggested that, because of Bessie's unstable life, she should send their daughter to family members.

"The only way is to have her away in a good healthy place with the proper people," he wrote her. "They need not adopt her, but it should look to the public that you are not the mother; you owe that to yourself and family."

Another former lover threatened to cut off her financial support if she didn't accept proper healthcare. "We never know what you are going to do as your word is not good," wrote the man. "One thing, young lady, you will line up with a hospital or some institution, or I will not have anything to say to you further."

Bessie had been sleeping at the Antlers Hotel in San Francisco for more than a week, receiving male callers there. But she was planning to check out on August 19, less than a week before her scalp and ear were discovered in El Cerrito. She asked her mother to meet her at a downtown corner in Oakland. They chatted about nothing really, and then Bessie fixed her fur-collared brown coat and collected her things. She had to dash off to an appointment, one that seemed a bit fishy to her mother. Bessie planned to meet a very prominent political figure in Contra Costa County, a sheriff named Frank Barnet.

"I won't be with him long, though," she told her mother. "I've got to see the doctor a little later."

Bessie kissed her goodbye and vanished. The next time Annie Ferguson saw her daughter, she was a skeleton lying on Oscar Heinrich's laboratory table.

The forensic scientist picked up pieces of Bessie's skull and ran his fingers along the jagged edges. Just days earlier he had used her bones to confirm a murder, and now he hoped it would help unmask her killer. He called Berkeley police chief August Vollmer to update him, even though it was being handled by Oakland police. Oscar typically discussed his theories with Vollmer, confiding in him even before he discussed the evidence with cops actually working the case.

When Oscar had first examined Bessie's ear, he suspected that the killer was a hunter or a meat butcher, someone skilled with a knife. But examining and assembling her bones changed his opinion—there was more finesse.

"The work is neatly done," Oscar wrote. "Seven starts were made with the saw before the operator was satisfied to cut through the skull on the primary separation."

She hadn't been hacked, but the cuts weren't precise. The killer wasn't a professional surgeon, but he *did* have some medical training. Oscar was building the profile of a murderer, much like he had in Father Heslin's case. Scotland Yard detectives had used criminal profiling almost forty years earlier to analyze Jack the Ripper. They believed he was likely a butcher or a physician, though a pair of London physicians later disagreed. They argued that the serial killer left too many gaping holes in his victims to have any real knife skills. Of course, Jack the Ripper died in anonymity—Oscar certainly hoped this case would fare better.

"The skill shown requires the candidate in this case to have had the ordinary training in dissection such as is given to physicians and dentists during their first year at a medical school," he told detectives.

The murderer used lime to age the remains, another mark of a scientific education and a calculating mind—this man wasn't a simple hunter. The killer likely had thick, dark brown hair based on strands found on the clothing that didn't match Bessie's hair. Oscar constructed a criminal profile based on those clues, and then he wanted to settle on a motive. If he could sort out *why* she died, it might point to the killer.

There were several possible scenarios. Oscar wondered if she had died during a botched abortion, because Bessie told her mother that she was planning to visit a doctor. But Oscar examined her torso and found no evidence of any procedure. The criminalist searched her discarded suitcase and discovered sanitary supplies. An unwanted pregnancy was unlikely.

Oscar thought about Charles Schwartz, the calculating chemist. He wondered if Bessie's killer had read about Oscar's work on that case. Charles Schwartz was clever, but he had left behind some crucial clues for Oscar.

Bessie Ferguson's killer had spent great effort concealing her remains in far-flung locations. He murdered her during one of the darkest nights of the month. He burned her flesh with lime—a meticulous cover-up that led investigators toward a disturbing conclusion. Perhaps he had killed before and might kill again.

Oscar Heinrich was happily married, there was no doubt. His wife, Marion, was steadfast, committed, and supportive—a marvelous mother and partner. She often accepted a minor role in his hectic public life; she had accepted that years ago. Oscar was constantly flattered by media attention—the name E. O. Heinrich could be found within most big crime stories on the West Coast by 1925.

But with those cases often came a curious situation that might have made an insecure wife wobble a bit. Heinrich's former girlfriends sometimes emerged, women who hoped to reconnect with the famous forensic scientist. In fact, the previous summer Oscar had lunched with a former flame named Louise in San Francisco, and he told his mother about it before her death. Unfortunately for Louise, it was a meal that left Oscar universally unimpressed with the women of his past.

"She tried to persuade me by her deportment and her interest in what I was doing and had accomplished in my own personal development that she was just as lovely as I used to think she was when we were young," Oscar wrote to his mother. "There is nothing which can become so utterly dead as a man's past love."

Oscar adored his wife; he had refused to join several professional organizations because their nighttime events had never included an invitation for wives. He was loyal, even as he joked about his priorities in life. "I love my pipe, my books, my wife, my children in about the order named," he wrote Kaiser. "Marion has been the lure for my spare time relaxations."

But they were aging and, at forty-three, the forensic scientist was candid about his marriage.

"Marion moves along in an orderly way," Oscar confided to his mother. "She is a fine mother to the boys and a helpful wife in every possible way. I notice her hair is beginning to turn a little gray. Not enough to cause any concern but yet enough to show that she as well as I are growing up."

As Oscar grew older he became most wistful about his youth and curious about the remainder of his life.

When he stepped inside the hunting lodge, Oscar could smell death. Gordon Rowe owned the small cabin; he was a San Francisco businessman who had been Bessie Ferguson's employer before he became her lover. He was missing, for now, but investigators were tracking him down. Police had found letters in Bessie's trunk, love notes that indicated Rowe's willingness to "support his child," as Oscar wrote in his field journal. Of course, there was no child, and police theorized that he might have discovered Bessie's scheme. Oscar knelt on the floor of the lodge, searching for blood. Later he would discover quite a lot, but not a drop was human blood—another dead end.

"I have inquired into his history and his training and into his behavior during the early days of this investigation," Oscar told investigators. "I have found nothing at all."

Oscar was discouraged—Gordon Rowe was unlikely to be the killer, and this case was proving to be exhausting. The criminalist reread his notes from his interview with Annie Ferguson, Bessie's mother. He noticed her especially terse language about another suspect, Alameda County sheriff Frank Barnet, the political figure who was on Bessie's appointment list the night she disappeared. Annie Ferguson believed that her daughter had told Barnet that she was pregnant. It appeared Bessie was running the same extortion scheme yet again. But this time, had she picked a less pliant victim?

"Says Barnett wanted her to go to a doctor, Bessie refused to go," wrote Oscar in his notes. "[Mother] claims Bessie was thoroughly disgusted with him."

The married sheriff was a strong suspect because of his public standing and criminal justice background. But the forensic scientist had taken soil samples from Barnet's home, and the organic material didn't match the sand he found at the crime scene. Oscar thought that Barnet might have been capable of killing a mistress, but this murder was too sophisticated for a simple sheriff without a partner.

"The skill shown in the dismemberment of the body to my mind excludes Barnett as the dismemberer," concluded Oscar. "The idea of two people being associated in this work one of whom has done such a complete job of dismembering at the behest of another seems to me too remote seriously to be entertained."

Oscar scanned his list of suspects, men who might have wanted Bessie Ferguson dead. The criminalist had dismissed Gordon Rowe and Sheriff Barnet, but there were five other names. One was a relative of the Fergusons who was a butcher, but he had provided a convincing alibi. Oscar was now left with four strong suspects: a successful physician, a respected surgeon, and two dentists. But soon he would focus on just one name—a curious man who had admired Bessie Ferguson from afar.

"Talking will fatigue men. The minute some important function of the body begins to get fatigued, it reacts upon your mind, upon your mental ability to resist and to maintain a lie."

The students stared back at their professor as he stood at a blackboard during his forensic science course at UC Berkeley. Since he began teaching eight years earlier, Oscar had developed a compelling lecture style that was energetic and engaging, especially when he explained how to interrogate a suspect.

"Where a man is willing to take the time to make his subject tired, he can get that result," he told the class. "Now that wasn't torture, but it comes as close to the third degree as anything we have."

After almost a decade of teaching, Oscar was one of the most popular instructors in the university's criminology department, a recent offshoot of the cop college that Oscar had helped create in 1916. His influence in the classroom was expanding because now he was educating criminal attorneys, social workers, bank officials, and other students of criminology.

Oscar held up photographic enlargements to illustrate how to detect forged signatures using the light and shade of handwritten letters. He pointed to the tiny tails on blood droplets during another class on bloodstains. These college courses provided the criminalist with a steady source of income, and by 1925, he was teaching hundreds of students.

John Boynton Kaiser was impressed with Oscar's progress as an educator, but he was also a bit alarmed by the quick education he had offered. He gently cautioned Oscar about giving students too much false confidence—good advice from a bright friend who had counseled plenty of blowhard researchers.

"A little knowledge is a dangerous thing," Kaiser warned. "One cannot be a finished psychiatrist, psychoanalyst and criminologist as well as a chemist and lawyer in five weeks."

The librarian was concerned that Oscar's students could claim to be experts, when in fact they were amateurs who might convince juries to convict on dubious evidence. And that's what soon began to happen, much to Oscar's dismay. Oscar described to Kaiser a frustrating encounter in court with a firearms "expert." "Typical of the policeman who has been sold in scientific investigation," he complained. "He seems to feel that with a microscope, a camera and a book or two that he can take the spotlight and hold it."

During the nascent era of forensic science development in the 1920s, just about anyone could market themselves as an expert witness—a dangerous trend in criminal justice and a potent threat to Oscar's financial future.

"The moral seems to be that if we show these fellows very much," he told Kaiser, "we will soon have as many ballistics experts as gunsmiths."

Oscar found himself increasingly at odds with other forensic specialists. Four years earlier, Oscar began writing Albert Osborn in New York, the country's foremost expert in document examination and forged document analysis. Their correspondence was cordial, but in 1925, Oscar became increasingly threatened, even paranoid of the other criminalist's motives.

"I have long held the opinion that in spite of his apparent altruism," Oscar wrote a friend, "there was in his makeup an element of selfishness and egotism."

Osborn had asked August Vollmer for a teaching position at UC Berkeley without talking with Oscar—a breach in courtesy. Osborn's classes would overlap with Oscar's course on forged documents, and his instruction would tread on Oscar's territory and threaten the job the criminalist needed so much.

"I do not yet know whether to look upon the action as hostile or merely arrogantly altruistic," he wondered to a friend.

Oscar Heinrich was compiling a long list of professional enemies, perhaps a bad decision for an expert who leaned heavily on his strong reputation and personal recommendations.

Oscar adjusted the oculars several times and then flipped the dial to increase the magnification. He peered through the lens at hundreds of blue-salt-and-pepper threads soaked with blood. He pivoted the eight-inch square of carpet, another clue from El Cerrito. It was early January 1926, almost six months after Bessie Ferguson's murder, and Oscar Heinrich was still trailing far behind her killer. But he felt as if he was getting closer. He had to be closer, because the woman's mother called on Oscar often enough to laden him with guilt. He was a parent, so he greatly sympathized. He squeezed a pair of metal tweezers and gently removed a curious substance.

"I found impressed in the fibres a tiny fragment of material which has been tentatively identified as dental cement of the kind used in making dental impressions," he wrote in his notes.

There was dentist modelling wax embedded in the carpet square intermingled within the murder victim's clothing. Oscar opened a container holding a gold crown that belonged to Bessie Ferguson. He looked at it under a magnifying glass first and then under a microscope and made a remarkable finding—tool markings made by a professional dentist.

"These dental fragments are in harmony with the training required to perform the dismemberment and particularly the dissection of the upper jaw from the skull," he wrote.

Oscar searched for his notes about Bessie's fatal injury, a hit to the top of her head that had caused multiple fractures. "Such an injury if it occurs in a dental office could easily be inflicted by a dentist upon a patient recumbent in a dental operating chair," he noted.

The killer was a dentist, a man with a medical background but limited surgery skills. Oscar reread his notes on the case. Ferguson's list of paramours and admirers included *two* dentists. He was inching closer to her killer. Oscar was confident that Bessie's murderer was a dentist—but which one of the two?

Oscar reviewed his list of suspects and underlined just one: Dr. C. C. Lee, a Chinese dentist who owned an abandoned building just across the street from Bessie's family home.

It had been ten months since Bessie Ferguson was bludgeoned and dismembered. It had been such a long wait for her mother . . . and for Oscar. He was desperate to close the case now—he refused to admit he had failed. He homed in on one man, a neighbor who fit the profile of her potential killer. The other suspects, a short list of men who had loved or lusted over Bessie, were cleared—but not without repercussions. Frank Barnet, Alameda County's sheriff for more than twenty years, lost his bid for reelection because of the Ferguson case.

"Dr. Lee," Oscar scribbled on his notepad.

The criminalist confirmed that Lee owned a garage along with an empty house right across the street from Bessie's parents'. He claimed to be a dentist, but there was no proof he had a medical education. But he did have direct contact with Bessie. Police found his shredded business card in her bag, and they discovered that his wife had often visited with her when Bessie returned to her parents' home.

Oscar was nearly obsessed with the case, even though he was embroiled in so many others. The criminalist hired a newspaper reporter from the *Oakland Tribune* to surveil Lee's home, but the journalist never spotted him. After almost a year of investigating, lab tests, and interviews, Oscar Heinrich finally admitted defeat in the case of Bessie Ferguson. Dr. C. C. Lee, the mysterious dentist, appeared to be a phantom, an elusive suspect who might have gotten away with murder.

"I am not prepared to say that he has committed this crime but I do believe that no investigation of this case will be complete without investigating this man," Oscar solemnly told investigators.

Throughout his forty-year career, there would be a handful of cases that nagged Oscar—Bessie Ferguson was one of the most painful. In 1927, two years after she was murdered, he was drawn into a case that seemed so familiar. A hiker traversing a trail in El Cerrito noticed a shallow grave filled with human bones. Police delivered another package to Oscar as they did with the Ferguson case—this one also contained a jawbone.

Oscar returned to El Cerrito and knelt near the burial spot, an area of the swamp that was very close to where Ferguson's ear and scalp were found. This victim was murdered, burned, and then buried. The two cases might have been connected, two people, both the prey of one killer who used the same dumping ground. Both investigations remained whodunits—Oscar Heinrich was never able to solve either case, so he had failed both victims. He would never forget Bessie Ferguson.

10.

Triggered: The Case of Marty Colwell's Gun

It is of the highest importance in the art of detection to be able to recognize, out of a number of facts, which are incidental and which vital. Otherwise your energy and attention must be dissipated instead of being concentrated.

—Arthur Conan Doyle,
The Adventure of the Reigate Squire, 1893

Martin Colwell stood at the door of the cottage on Pennsylvania Street. He reeked of alcohol, as usual—it had been his vice for years. The police in Vallejo, California, had written his name on their offender blotter for different reasons, most notably assault and burglary. The state prison had hosted him for three terms since his youth. Martin "Marty" Colwell was a mean drunk, a ruffian who rankled police and issued threats to intimidate just about everyone else.

And now he was furious over yet another betrayal. As he pounded on the cottage's door, he held the .38-caliber revolver in his right hand. He focused his eyes ahead—he readied himself.

It was December 19, 1925, just days before Christmas. The door opened, and John McCarthy, Colwell's former employer, staggered into the doorway. Blood was sprayed across his shirt. Colwell heard moaning, saw McCarthy's wound, turned, and ran. McCarthy lurched onto his front stoop, begging for help. Police responded quickly and tried to save his life, but it was too late. McCarthy, a loved and respected small-business owner, muttered some final words to investigators, a simple phrase that seemed likely to condemn his killer.

"I fired Colwell."

It seemed like an open-and-shut case, but Colwell quickly made a savvy decision: he hired an impressive defense team. The district attorney, panicked, phoned Oscar Heinrich. Colwell would certainly murder again if he wasn't convicted, the DA feared.

The press ignored the murder. This wasn't a beguiling story about a devious chemist or a deranged baker, but Colwell's trial was crucial because it offered Oscar something he desperately needed—redemption.

The remainder of the 1920s had produced a string of cases that challenged Oscar's scientific acumen and his investigative skills. He was one of the first to use ultraviolet light to reveal blood. Oscar used deductive reason and poison tests to prove murder by strychnine when toxicology was still in its infancy. He honed his technique in blood-pattern analysis after the shooting death of a wealthy wife.

As America approached 1930, Oscar's forensic tool belt was fortified: ballistics, botany, toxicology, chemistry, document analysis, and many other disciplines within the field of criminology were being developed by brilliant pioneers, including Oscar. It was also a decade of incredible growth, and there was still more to come. Blood typing, voice printing, trace blood detection using luminol, and semen tests were more than a decade away. But there were growing pains as scientists battled mistrustful detectives, skeptical prosecutors, overwhelmed juries, and undermining experts. Oscar sometimes felt he was battling a war on two fronts: first to develop the techniques in the first place, and then to bring the public and law enforcement along with him to understand and trust these new scientific advances.

But it was a war he was winning. He didn't triumph in every case, but slowly over many years Oscar's reputation was being bolstered with each successful prosecution. Even better, his finances were finally stabilizing near the end of the 1920s. The period of national prosperity in Oscar's life—and in America as a whole—seemed as though it would never end.

But of course, all good things *must* end.

The silver object came into focus, then rose up and down, blurring and clearing with every turn of his dial. He kept a notebook and a pencil nearby, stopping occasionally to record scribblings of long numbers and small drawings that only he could put into context.

Wall bullet: 144.69 grams 0. 3546
9.3760 grams 9.01mm

After several minutes Oscar settled on the appropriate magnification. His eyes widened at the simple design on a single bullet pulled from a wall inside John McCarthy's cottage. Oscar counted each of the ridges on the bullet, the numerous minuscule markings that he could detect only through the lens of his microscope. The tiny faded gray lines on the discharged slug were innocuous to a layman, but they would serve as a linchpin in this murder case.

Oscar looked over his notes. Witnesses at a nearby ice plant heard a gunshot before McCarthy staggered from his home and pleaded for help. The fatal injury had been a "through-and-through," a police phrase for when a bullet passes through a body.

When they arrested fifty-nine-year-old Martin Colwell, he was armed with a .38-caliber, five-chambered hammerless Smith & Wesson revolver—the same caliber of gun used to shoot his former boss. Police discovered three bullets in Colwell's coat pocket and a box of bullets in his boat. Colwell denied killing McCarthy, blaming amnesia from too much liquor.

"I'll get you all," Colwell screamed to police. "Just wait until I get out of here."

Oscar was enamored of ballistics, and he marveled at the power of firearms. He vied to become a preeminent ballistics expert, using guides like John H. Fischer, Calvin Goddard, Charles Waite, and Philip Gravelle, a cohort who formed the Bureau of Forensic Ballistics in New York City in 1924.

Forensic ballistics dated back to 1835, when a member of the Bow Street Runners, the precursors to Scotland Yard detectives, spotted irregularities in a bullet. Firearms in early America were custom-crafted by individual gunsmiths, so each part, including the barrel, screw, and bullet, was unique. The markings on the spent bullets were also exclusive to that specific gun, so experts could connect a weapon to the deadly bullet. They squinted through monocles at lead, searching for scratches left by the gun's firing pin. They noted the distinct spiral marks that were engraved by the grooves cut into the interior of the rifled barrel.

When gunmakers began to mass-produce weapons in the nineteenth century, experts turned to microscopes to spot more subtle irregularities. Weapons still left behind distinct markings—they were just more difficult to trace.

In 1926, Oscar Heinrich was one of the first professional criminalists to use a new invention by chemist Philip Gravelle, the comparison microscope—one of the greatest contributions to the development of firearms identification. When analysts used two separate microscopes to compare a pair of bullets, they had to trust their memories as they moved back and forth. The comparison microscope allowed its user to compare two images at the same time—offering scientific precision along with credibility in court.

Oscar hovered over a pair of Bausch & Lomb lenses equipped with two forty-eight-millimeter objectives—two microscopes with a pair of slides that were sitting side by side connected by an optical bridge. The device used mirrors and prisms to direct the light from each microscope to a common viewing area; a pair of oculars allowed the user to examine the two samples simultaneously, with a split screen. The first lens focused on the lethal bullet, the one that had passed through McCarthy's body and then buried itself in his wall. Oscar slid a second bullet under the other lens, one that police discovered in the suspect's coat pocket.

Oscar pointed Colwell's gun, the .38 revolver that likely killed McCarthy, at a barrel of paraffin wax and fired once. A bullet exploded from the gun, embedding into the wax. Oscar waited. Soon he carefully cut it from the thick material and soaked it in gasoline to dissolve the paraffin. Now the bullet found in Colwell's coat would show the gun's unique markings. If the size, shape, and direction of the twist of the grooves from both bullets matched, Colwell would likely be convicted solely based on forensics—case closed, Oscar thought. But even he should have realized that many of his cases never closed that easily. Oscar believed that this investigation would hinge on the scoring of the ridges, also called rifling marks, on the deadly bullet.

"These ridges," Oscar explained, "make their imprint on any projectile as clearly as a finger can leave its print.

"The raised ridges were called 'lands,'" he explained, "while the pattern's shallow, small depressions were labeled 'grooves.'"

Investigators stared at him, clearly confused. Time to change tactics. Oscar imagined they were his college students, naïve about most things inside a lab.

"The riflings are made by the cutting edge of a steel bar that is shoved in and pulled out of a barrel many times—fifty times in some cases and as many as a hundred in others," he explained. "The process leaves the bottom of the grooves serrated, so that no two such marks are alike."

The detectives seemed to understand then, sort of. But they certainly were pleased with one bit of good news. The "bullet fingerprints," as Oscar called them, were conclusive—Martin Colwell used his gun to kill John McCarthy.

As Oscar sat alone in his lab, he grew uneasy. The detectives seemed dazed, just like the jury in Fatty Arbuckle's case. His testimony might have been too technical once again, a complaint that jurors repeated throughout his career. In fact, just a few months earlier, another jury had acquitted a murder suspect despite Oscar's conclusive ballistics evidence. The panel in that case was addled by his exhaustive explanation of rifling marks. He hadn't learned—once again he had touted his scientific pedigree at the expense of clarity . . . and lost the case.

He refused to allow another killer to escape prison, so Oscar prepared very differently for this trial; he was determined to *show* the jury why he could prove that Martin Colwell was a murderer—he just had to find a way. And then he had an idea.

In his lab, the forensic scientist hauled his favorite camera to his desk, a Zeiss Ikon Prontor, and pointed it at the comparison microscope's dual optics.

"Caught image on ground glass in focus circa 140mm," he wrote in pen. "Adjusted for center by looking through lens from rear, sighting on crosshairs on focusing glass."

It was so simple. With a few snaps of his shutter Oscar Heinrich shaped the future of forensic ballistics by pioneering a technique that would solve decades worth of cases.

The courtroom was overfilled when the bailiff cautioned spectators to stay quiet; the judge leaned over his desk as Oscar sat in the witness chair and explained how he used nearly invisible ridges and grooves to catch a killer.

"There were four panels on this bullet which were undamaged and sufficiently clear to examine microscopically," he told the jury. "These four panels proved to be in every detail similar with the four corresponding panels on the test bullet under examination."

Oscar pulled out from a folder a black-and-white photograph showing two bullets, side by side and separated by a thin black line.

"I then proceeded to make a photomicrograph of what I found with the same instrument just as the eye saw it."

The assistant district attorney noticed jurors with quizzical looks. "What is that, professor?"

Oscar ignored the silly job title reference and continued. "A photomicrograph is an enlarged picture made by substituting the microscope for the ordinary photographic lens and making a picture through the microscope with a combination of magnifying lenses," Oscar replied. "If I might have a blackboard I would like to explain this more thoroughly."

Oscar stood up and handed copies of the photo to both sets of attorneys along with the jury before walking over to a portable blackboard. He drew a circle with chalk and sketched a diagram. Oscar hoped to offer the jury a simple explanation for *how* this new technique worked.

"The photographic print shows a circle. Across this circle is a black line. That is the dividing line between the two microscopes," he said. "I have two microscopes hitched together and I am looking at two bullets."

He interpreted forensic ballistics, the importance of those tiny, crucial markings left on the pair of bullets.

"Now, in the gun there is a series of runners within the barrel, evenly placed apart, these runners have made grooves on the bullet," he said. "There are five grooves and then five high points on the bullets between the grooves."

Oscar showed the jury how the grooves on the test bullet from Colwell's pocket matched the grooves on the lethal bullet dug out of McCarthy's wall, the one that went through his body.

"They match up line for line," he told the jury. "These are the marks on the bullet made by the defects of the barrel."

No two guns would contain the same defects inside their barrels, he assured the jury, so no two guns would produce identical grooves on bullets. Colwell's defense attorney had remained calm through most of

Oscar's testimony, but his mood quickly changed as jurors thumbed through the photos.

"I will make a general objection to the introduction of any of these pictures upon the ground that it is incompetent, irrelevant, immaterial, and the proper foundation had not been laid for their introduction," Arthur Lindauer bellowed.

"The objection will be overruled," replied the judge.

Oscar grinned, but not for very long. Colwell's defense team attacked every bit of his testimony, including how he dismantled and examined Colwell's gun, how he cleaned the bullets, and even the way he calculated the size of the bullets.

"Will you compute three-eighths of an inch?" demanded Lindauer. "It would be point 38, wouldn't it?"

"No, it is point 35," replied Oscar calmly. "The reading point 38, that is different from three-eighths."

"I understand," said Lindauer.

"Well I want to know if the jury understands what I am talking about," Oscar snapped.

"I wonder if the jury does," replied Lindauer snidely.

Oscar was being undermined over science, and he was incensed. After that curt exchange, the prosecutor dismissed him. As the criminalist sat down, he scanned the defense's witness list—his old rival Chauncey McGovern, the defense expert who had challenged Oscar during many cases, including Father Heslin's murder four years earlier.

Oscar was miffed—now McGovern claimed to be trained in forensic ballistics? It seemed ridiculous. His first appearance at trial as a firearms expert was just a few weeks earlier, during a murder case that still enraged Oscar.

McGovern had made some incorrect claims on the stand, lies that contradicted Oscar's conclusions and flummoxed the jury. The murderer had been released.

And now this trial was matching the two bitter adversaries in court once again. McGovern was about to accuse Oscar Heinrich, the nation's most well-known forensic scientist, of fraud. Oscar sat close to the district attorney and glared at McGovern. He seethed as the fifty-three-year-old sat in the witness chair. McGovern held up Oscar's photos of the test bullet and the lethal bullet, photographed side by side, and stared at the jury.

"Absolutely a physical impossibility," scoffed McGovern to the jury. "The microscope would not let you take the picture. It is an absolute physical impossibility to place those bullets together into some shape that you can make a photomicrograph showing the lines alongside. You have to make a picture of this, and a picture of that, and paste the pictures together."

The jurors listened as McGovern accused Oscar of taking separate photos, trimming them carefully, and photographing them together as a composite—a magic trick, a devious sleight of hand. And if Oscar had fabricated that experiment, *none* of his testimony could be trusted, concluded McGovern. One juror was confused and raised his hand.

"Suppose you take a photograph of one and a photograph of the other, how are you going to take that line as to make them correspond?" he asked.

"You would cut off part of one picture and cut off part of the other picture and put the print parts together, like Mr. Heinrich did here," McGovern replied.

The jury still seemed confused about how that one small, perfect black line separating the two bullets could result from two photos that were smashed together. Oscar had offered them a simple explanation earlier: the line was the intersection of the two prisms from the two lenses of the comparison microscope.

"I just wanted to know how you get the lines on the picture?" another exasperated juror asked McGovern.

"That is the only way, you cannot get them through the microscope," replied McGovern, "you have to take them with two pictures."

McGovern seemed resolute—a forensic scientist simply could not photograph two bullets under a comparison microscope. Yes, he admitted, he would be able to photograph two bullets under a normal microscope with just one lens, but three-fourths of each bullet would be cut off by the eyepiece. He absolutely refused to believe that Oscar's photos were authentic.

As the prosecutor pummeled McGovern over his lack of experience with ballistics, the expert became defensive and then apologetic.

"I had four hours sleep last night," McGovern explained to the judge. "I am at your disposal, if it takes us until hell freezes over, very gladly, to use a popular term."

Jurors gasped, and Oscar looked surprised. McGovern shrank in his seat and looked over at the judge.

"Now, I don't think we need any expression of that kind," snapped prosecutor Leo Dunnell.

"Listen, Mr. Expert," warned the judge, "can't you express yourself in more respectful words, and indulge in a language that is fit?"

"I stand corrected, Your Honor," he replied.

The assistant district attorney quickly thought for a moment and whispered to Oscar, who smiled. They would set a trap. He asked McGovern if he thought Oscar Heinrich could replicate his experiment in court in less than five minutes.

"His answer was that I couldn't do it in five minutes or two hours," Oscar said to a friend.

With that haughty declaration, McGovern fell for the ploy. Oscar returned home to retrieve his laboratory equipment. Soon he was carefully setting up his comparison microscope and a camera on a long table in front of the jury.

"I noticed some of the jurymen look at their watches," Oscar remembered. "I went right ahead and had the setup complete in four minutes."

Oscar placed the two bullets under his microscope, and soon each juror stepped forward and peered through the lenses. Colwell's defense attorney quickly objected.

"Now, if Your Honor please," pleaded Lindauer, "not for the purpose of looking through to see the bullets; the purposed is to explain how he took the picture; that is all."

The judge overruled the objection and issued a simple request to jurors as they stood up: "Be careful not to jolt the table." Each man bent over the microscope as Oscar issued instructions.

"The eye should be brought right close to the lens," Oscar said. "It is not very safe to put your weight on the table."

They nodded.

"I found that, in working with this microscope and studying it for its possibilities, that if I could put my camera lens in this position, instead of my eye," Oscar explained as they filed past, "that I would get a photograph on my negative. Gentlemen, that is what I did in taking this photograph."

It was clear to each juror that the grooves and ridges on the pair of bullets matched—they were fired from the same gun. But one juror looked up, still doubtful of the funny photographs and the suspicious science behind them.

"I'd like to see him actually shoot that picture here again while we're all watching him do it," said the juror, "and I'd like to look at the picture after it's been developed."

Oscar smiled and agreed. He strode over to his setup and snapped the photos. A commercial photographer was summoned and traveled with him to a nearby darkroom and waited with him for the pictures to develop. Later that afternoon, Oscar stood before the jury with the new photographs—they were identical to the ones he had previously taken in his lab. McGovern fumed as Oscar remained characteristically stoic.

The jury retired, and after days of deliberating, the panel emerged . . . with no decision. The jurors were deadlocked over the scientific strife between Oscar Heinrich and Chauncey McGovern. The judge declared a mistrial. It was so aggravating to Oscar—how could they not believe him, after all his demonstrations?

The district attorney launched a retrial in May with the same evidence and the same experts, including Oscar and McGovern. Oscar readied himself for dire news when the jury returned in less than two hours—it might be another hung jury. But it was good news, finally. Guilty. They recommended life in prison for the man who gunned down his former boss.

Oscar's ingenious technique using a camera and a microscope established an international legal precedent. It inspired refinement of the equipment so that scientists today can offer more efficient, accurate evidence.

But like other forensic tools, firearms identification can still be challenged successfully in court because of a lack of scientific study. The 2009 landmark report from the National Academy of Sciences concluded that forensic ballistics can be valuable during criminal investigations, but the panel cautioned that those tests should never be the only source of evidence.

"The committee agrees that class characteristics are helpful in narrowing the pool of tools that may have left a distinctive mark," read the report. "Individual patterns from manufacture or from wear might, in some cases, be distinctive enough to suggest one particular source, but additional studies should be performed to make the process of individualization more precise and repeatable."

As with most other forensic disciplines, the legal system must demand more tests and more accountability from ballistics analysts. But in 1926, Oscar's success in the murder trial of Martin Colwell was a massive step forward for forensic ballistics. Firearms pioneer Calvin Goddard begged Oscar to send him the photos from the case to illustrate how the comparison microscope could be used in court. The following year, Goddard would use the same device in the famous Sacco and Vanzetti case. Oscar continued to refine his technique, offering to collaborate with other ballistics experts.

"I have myself added a comparison eye-piece to my outfit," he told one expert. "If you will bring your monocular along we can see what can be done toward uniting the images obtained through microscopes of different makes."

Microscope manufacturers hired Oscar as a consultant to help adapt their own equipment for legal cases. Attorneys around the country requested him.

Colwell's second trial served as an epiphany for Oscar—jurors could be capricious regardless of their jobs, their education, or their religions. But he felt optimistic because he had finally sorted out how to connect with a jury. If he couldn't seem to explain it to them, then he would show them in every way possible. Jurors were reticent to convict on evidence based on strange science. Now he understood.

The scars left by Fatty Arbuckle's case seemed to be finally fading. And Martin Colwell's trial offered Oscar a belated but gratifying victory over Chauncey McGovern. They would continue to face each other in court for another seven years, but "America's Sherlock Holmes" would always be the favored expert.

"Happy birthday, old top, and may your shadow never grow less!"

Oscar greeted his closest friend, John Boynton Kaiser, at the end of 1926 in his annual Christmas letter, which was a customary review of his life. The criminalist was still delighting in the attention from other scientists in the wake of Martin Colwell's murder trial. His business had swelled, which was welcomed news, but it also proved to be a hinderance to his family life.

"I get my greatest joy in life out of my family," he wrote Kaiser. "I see them practically but once a day. This is at dinnertime in the evening."

And that trend would continue for two years as he traveled almost constantly.

"I have been too busy. During the last sixty days I have worked in Colorado, Kansas, Utah, Nevada, Oregon and miscellaneous spots in California," he wrote a friend. "I find that I haven't the endurance that I had twenty or even ten years ago."

He was trying to limit his time at work to just eleven hours a day, a nearly impossible goal, and now he felt even more financial pressure. When Theodore turned eighteen in 1928, Oscar considered his request to spend six months traveling throughout Europe. His elder son hoped to develop a keen eye for artwork and architecture while fulfilling a language requirement for his upcoming graduate studies at the University of Cambridge in England. Oscar scoured his many financial logs and sketched out numerous budgets before reluctantly agreeing. Theo spent months traveling England, Germany, Italy, and France. He sailed through the Panama Canal and studied inside the world's most famous museums. His trips drained Oscar of thousands of dollars, and the fiscal strain was affecting his business.

"I am returning several book lists that have been on my desk, I wanted to possess them," he wrote Kaiser. "I have heard from Theodore, have arranged to bring him home and am now temporarily broke. You can draw your own conclusions."

Mortimer, Oscar's younger son, was now fourteen, and both boys were becoming expensive dependents.

"Each had to be shoved through the hospital for an appendicitis operation. Then they had to be supplied with vacations to recuperate," he told a friend. "Next Mortimer became tooth conscious and now is having teeth fixed. On the other hand, Theodore provided a hot finish to his vacation by burning up his Studebaker car."

But Oscar also greatly admired his maturing son, because Theodore was developing into a prolific writer, like his dad, by publishing articles about gardening in high-profile magazines.

"Perhaps I was a little sentimental about it," he wrote a friend, "but the boy had just sold another article to *House Beautiful* on his own initiative and merit."

Eventually Theo would find more success in national magazines than his father. He was a bright student with a seemingly charmed life—a gift bestowed by a parent who had little growing up.

"I shall be with you in spirit, proud that you have a purpose to become useful to the world, ready with my meager reserve of strength to support you," he wrote Theo. "If you succeed I too shall do so vicariously. If you fail to come back to me, I shall not grumble. I shall do as we always have done—mop up, reorganize and start over."

But his son continued to spend money without submitting financial reports, despite Oscar's repeated frantic requests. He had even threatened to withhold Theo's monthly allowance.

"If you don't pay attention I shall delay getting off a remittance until you take the time to satisfy me," Oscar wrote.

For years he worked too much with no rest, drowning under financial debt. And soon pride in his sons slowly developed into resentment. "I cannot avoid feeling somewhat uneasy about the time you have given over to roistering on the Mediterranean to the seeming abandonment of parts of what I understood to be your program," Oscar warned Theo. "California is now entering the effects of the depression. What you must do to cooperate is to study economics during your spare time and prepare yourself to skid along the bottom with the rest of us."

Oscar was mildly encouraged by the hint from Theo that he considered being an engineer, like his father once had been, but Oscar was also concerned about his son's sense of responsibility, his lack of urgency over money. Theo didn't seem to understand how the fear of financial failure had deeply affected his father—or his grandfather August Heinrich. Oscar tried to explain to Theo how his own father's incompetence had almost doomed the teenager before his life had really begun.

"Your grandfather was creative but without training, and died a poor man leaving your father at your present age to win fame and fortune unaided with advice or guidance," Oscar wrote, "and with the added handicap of having to apologize for some of your grandfather's blunderings."

Oscar was exhausted from obsessing over personal budgets and an erratic income. And he fretted now over Theo's future.

"If I should die tomorrow there would be no support for you and to carry out your present ambitions you would find it necessary to work your way as I did," he warned Theo. "I am approaching that age where I begin to feel my energies curtailing."

Oscar often felt conflicted over his children, and it was troubling, especially as he grew older. He wanted to offer his sons a privileged life, a chance to pursue their own satisfying, lucrative careers. He pressured them to succeed, but he also warmly welcomed them home; he complimented both sons often. They strived to please him, an often-futile effort. Oscar was constantly running from his father's demons, abused by his own insecurities. He confided in August Vollmer that somedays he felt simply beaten.

"I must overcome the fear complex which I acquired in the early days of my professional career when I was being chased by the sheriff every ninety days or so to yield unto my creditors what I had hoped might be bestowed upon my hungry dependents," Oscar told Vollmer. "It was then I learned to live each day so that I could tell the banker to go to hell."

By his late forties, Oscar seemed unsatisfied with his own life despite incredible professional success. He spent afternoons gazing through the large picture windows of his laboratory overlooking the Golden Gate.

"I am persistently and continually beset with the desire to see for myself what the ships meet with beyond the horizon to the west," he wrote a friend. "Someday I am going to follow the lure of the sun and trek westward-ho."

But he wasn't destined for that kind of life. And in less than three years, America and Oscar Heinrich would struggle with so much more than simply lost dreams.

In 1929, a disastrous string of events created the country's worst economic crisis in its history. In March, the stock market had a correction, stunning anyone who had invested money they borrowed on margin from their stockbrokers. Millions of people went bankrupt, and nearly 25 percent of the labor force was jobless by 1933. Banks closed as Americans rushed to withdraw their money.

In October, the Black Thursday stock market crash sent stock prices plummeting by almost 25 percent. It would cost investors $30 billion, which was almost the expense of World War I. The profits of the Roaring Twenties were wiped away. It would become the longest and most severe depression in the history of the industrialized Western world.

Crime rates skyrocketed while suicide rates rose. In 1931, a damning government report condemned Prohibition laws for its negative effects on American society, leading to its repeal. August Vollmer was one of the authors of the Wickersham Report, which castigated police for failing "to detect and arrest criminals guilty of the many murders." Those accusations stemmed from several high-profile organized crime cases, including the 1929 St. Valentine's Day Massacre in Chicago, when Al Capone's gang, dressed as policemen, gunned down seven men. Ballistics expert Calvin Goddard—the inventor of the comparison microscope—used bullet markings to prove that police had not been responsible.

Goddard was soon offered funding to start the first public, independent forensic lab at Northwestern University's law school; the University of Chicago was close behind, and then Columbia University wanted a similar laboratory. Forensic experts were creating partnerships at real labs funded by wealthy universities. And police departments were finally admitting that forensics *could* solve crime.

That year, Goddard suggested that Oscar apply to teach at his lab at Northwestern. He would be joined by Luke May, a self-taught criminalist and sought-after expert. Chauncey McGovern was no longer problematic for Oscar after being publicly chastised in the Colwell case. But other experts would replace McGovern on Oscar's list of adversaries. Luke May's lack of education, flamboyant style, and arrogance absolutely riled Oscar.

"May is an egotist of the first order," he wrote. "His love for publicity is almost a disease."

And Oscar made public his acerbic opinion of the criminalist when he reviewed May's newest book, *Crime's Nemesis,* for a literary journal.

"The chapters have a surplus of fact and a deficit of thought and sentiment," wrote Oscar. "The author has done too little to help the public to comprehend and to realize more clearly and fully ideas on scientific criminal investigation that they now grasp imperfectly."

Calvin Goddard, Oscar feared, would inadvertently staff the lab at Northwestern with charlatans, fakes of the first order. And they would destroy the credibility of forensic science.

"Their tendency to over-emphasize their accomplishments is the one thing that keeps them from achieving the highest success," Oscar told a fellow forensic scientist. "They do not need to be feared as competitors any more than *True Detective Stories* need to be feared as a competitor of the *Atlantic Monthly.*"

The competition for these faculty positions was vigorous—most criminalists would have coveted a cushy position at a prestigious university like Northwestern. Oscar wasn't eager to teach at Northwestern because he was considering another position at the University of Chicago.

"The University of Chicago has the inside track," he wrote a friend. "I believe their work will be of the most constructive character."

Soon a decision was made—and not in his favor. Oscar had never felt jealous of a close friend . . . until now. The University of Chicago selected August Vollmer to teach there in 1929. Ultimately the dean thought the "father of modern policing" was the better candidate than the criminalist. Oscar felt wounded and then bitter. He believed Vollmer didn't have the forensic chops for a position at the college.

"I am the only man in these United States eligible to fill out that faculty," he fumed to a friend.

But he didn't let that bruising derail him. Oscar's confidence and arrogance continued to build with each case he solved, with every newspaper article that mentioned his name. His hubris, though mostly in check in public, was hardening even regarding experts he had once respected.

"I do not entirely agree with your assertion that any half-witted mechanic could build a triple bullet holder," Oscar told one ballistics expert. "But I will concede that said half-witted mechanic readily can do so if a Heinrich stands over him."

By age fifty, Oscar's ego seemed healthy, but his stamina was weakening. He could no longer work twelve-hour days. He spent months recovering from acute colitis, an inflammation of the colon made worse by stress. He habitually smoked, ending each evening with a "night-night pipe."

In the early 1930s, Oscar's taxes increased while the fees of all forensic experts were drastically reduced. He had solved more than a thousand cases by this time, but his jobs were cut in half. Police chiefs and attorneys could no longer afford him. Oscar's experience was akin to the collective investigative skills of an entire police force, but now he was underemployed and saddened from missing his twentieth wedding anniversary.

In less than a year, Oscar would work on a huge, lucrative case that would draw him into the spotlight and controversy once again. His contentious theory about the death of an affluent wife on the campus of Stanford University would test the legal system . . . and his skills as a forensic scientist. Oscar Heinrich would usher a fledgling forensic technique into legitimacy, perhaps to the detriment of justice.

11.

Damned:
The Case of Allene Lamson's Bath, Part II

I had already come to the conclusion, since there were no signs of a struggle, that the blood which covered the floor had burst from the murderer's nose in his excitement. I could perceive that the track of blood coincided with the track of his feet.

—Arthur Conan Doyle,
A Study in Scarlet, 1887

She squirmed in her chair as she watched the back of his head. Her brown eyes drooped at the corners—she looked so much like her father. Her hair parted to the right, wavy at the top but falling in curls. It was Monday, June 5, 1933, and Allene Genevieve Lamson, better known as Bebe, was puzzled. Her father sat so close by, just a few feet away, but she wasn't allowed to hug him. He seemed agitated. She called out, "Daddy!"

It had been almost a week since her mother last held her. Bebe tried to sit calmly, but she couldn't resist wiggling. There were more than one hundred men in dark suits and women in fancy dresses all sitting shoulder to shoulder, most struggling to stay hushed. More than a thousand people crammed into the entryway of the courthouse, and even more stood in the square near the county jail. A confluence of excited conversations erupted behind her—the roar seemed almost unbearable, at least to a toddler.

She rested her hand on the thick wooden railing as the judge, draped in black, surveyed his courtroom. Bebe's white, lacy dress studded with small beads would have been perfect church attire on another day. She nearly glowed atop her grandmother's dark suit and stylish black hat—more of a funeral dress than formal outfit.

"Yes," her father muttered several times to the judge. Bebe couldn't understand the questions. She squirmed again. Nearby her sat her middle-aged aunts, Dr. Margaret Lamson and Hazel Thoits. Bebe had been living with Margaret for the past week.

Quickly her father stood up and slid out from behind the large desk. A man in a black uniform placed a hand on his back before they vanished behind a door. Bebe glanced at her grandmother. The Santa Clara County Superior Courthouse was a bewildering, sad place for Bebe Lamson during David Lamson's arraignment—and it would only become more troubling, as her father would soon face a jury for beating her mother to death.

By 1933, Oscar Heinrich's business began to suffer from the impact of the Great Depression, just like the rest of the country, so he redesigned his forensic science firm. He expanded his foreign business by testifying against Germany in a war sabotage case—the American government proved to be a more reliable client than local prosecutors and defense attorneys. He began authenticating artwork in Europe. But these new enterprises still couldn't offset his dramatic loss in income.

Oscar complained that 50 percent of banks were frozen in Nevada, a major source of his business. There appeared to be a moratorium on prosecutors there to hire any expert, surely good news for the state's criminals. Oscar's California business dropped by one-third, and he had difficulty convincing clients to pay him quickly. All those circumstances, along with his elder son's coursework in England, were threatening to bankrupt the Heinrich family.

The year before, Theodore had traveled to Chicago, New York, Boston, Philadelphia, and other American cities as part of his studies at the University of Cambridge. The twenty-two-year-old sent his father a humble letter that included a list of travel costs.

"I've just been figuring expenses and have come to the conclusion that France is still a comparatively expensive country," Theo wrote Oscar. "I don't understand why the money has gone so fast, for I certainly haven't been extravagant."

Oscar deeply loved his son, but he offered little tolerance for money woes after years of being humbled by crippling debt.

"I fear you are still inclined to live more in a world of make believe than of fact," he lectured Theo.

Oscar was traveling more than ever, and he was concerned for his wife's safety as she stayed alone at their house in Berkeley. He requested routine police protection, saying he was afraid of disgruntled criminals.

"Say to Mort that I think police protection during my absence will be more efficient with the guard inside the house after bed-time than out," he wrote Marion.

Oscar Heinrich felt weak from worry by the beginning of the Lamson investigation. He was fifty-two years old by then and had less stamina than ever. His old adversaries had finally retreated—the crochety old handwriting expert Carl Eisenschimmel had died the year before, while the ever-troublesome Chauncey McGovern would die the following year. But their mutual vitriol for Oscar would be dwarfed by another nemesis, a German physician with charisma who threatened to derail Oscar's jury in the most crucial case of his life.

Inside his home laboratory the forensic scientist stared at the skull and then at its measurements. He scribbled notes. He adjusted the strings wrapped around it. After nearly two decades of trial work, he knew that preparation would be his most potent weapon in court.

There was chalk residue on his arm from a giant framed blackboard hanging in his laboratory in Berkeley. There was a crude sketch of a skull on the slate, an oval with no eye sockets or any discernible features except for two squiggles for ears. He drew four lines that swirled out from the center of the head, labeled A, B, C, and D, all meeting in the middle of the skull. Other wayward lines represented the smaller fractures that erupted right before Allene Lamson died. The pathology report spelled it out: there was one large fracture and three smaller ones. The prosecutor believed that David Lamson had struck his wife on the back of the head with a heavy weapon—four hits resulting in four fractures.

On Oscar's shelf sat a new book, courtesy of John Boynton Kaiser: *Simplified Blood Chemistry as Practiced with the Ettman Blood Chemistry Set,* a how-to manual for calculating and then interpreting bloodstain pattern analysis evidence. Oscar scanned the notes from his numerous experiments. They were all satisfying exercises, but he despised the man who had required them to be completed.

Before the trial began at the end of August 1933, Oscar had agreed to conduct joint forensic tests with Santa Clara County's pathologist, Dr. Frederick Proescher. Attorneys wanted to parcel out only small bits of evidence for testing. The tenuous concession fostered a competitive and testy dynamic between a pair of star witnesses who were angling for attention on the witness stand—both men with notable credentials and immense egos.

The fifty-five-year-old German had medical expertise at crime scenes and inside his research lab. He was the most qualified of all Oscar's challengers in court over the past thirty years. Now Dr. Proescher's name would be added to the list of Oscar's rivals, setting off a public feud between a medical physician and a forensic scientist in court.

The more Oscar studied his notes on the crime scene, the more alarmed he became. On the day of Allene's death, Dr. Proescher pointed to stains inside the closet of Bebe's nursery where the little girl's nurse kept her clothes. Blood, he had assured an officer with him, but Oscar tested the substance himself, and it was negative for blood. It was likely crayon, Dr. Proescher later admitted. The pathologist noted that David's shirt, one that he had not worn that morning, also had spots of blood.

"Yes, there was some blood on that too," the physician told the district attorney.

But Oscar's tests concluded those spots were also negative. There are blood droplets on the kitchen door, Dr. Proescher insisted to the prosecutor, but Oscar later confirmed they were spots of varnish. His incompetence was maddening to the forensic scientist—Proscher was a physician who didn't seem to understand chemistry or biology.

The morning of Allene's death, Palo Alto cops sifted through the ashes of David's bonfire in the garden. They examined charred curtain rods, pieces of garden hose, a small spade, and even some Chinese coins. If an item was rubbish, David Lamson had apparently tried to burn it that weekend. Dr. Proescher phoned Oscar and told him that a piece of cloth from the fire repeatedly tested positive for blood.

"I made more than one test," Dr. Proescher said. "I tried hematoporphryin, hemochromogen and benzidin. All positive."

These were all valid chemical tests using substances that changed color when mixed with hemoglobin, the protein in red blood cells. Proescher insisted that all three tests confirmed blood on the cloth, but Oscar was skeptical. Come to Berkeley, he replied. In June 1933, Dr. Proescher arrived at Oscar's lab carrying several samples of evidence from the bonfire. The pathologist demanded vindication—the cloth from the bonfire had tested positive for blood in his own lab. Oscar studied him as he handled a pair of scissors.

"Dr. Proescher took the sample and cut off about two-thirds of it, powdered it and made the benzidin test," recalled Oscar. "That was negative. From the residue of the powder I made a Leucomalachite Green test and that was also negative."

Now it was Oscar who was vindicated, because two tests had confirmed his findings—no blood. Dr. Proescher seemed to have a vendetta against David Lamson, Oscar thought.

After Dr. Proescher left Berkeley, Oscar returned to his own tests, and he made a crucial discovery using bloodstain pattern analysis. There were pools of blood on the floor, "passive stains" that typically resulted from gravity acting on an injured body; blood flows and drops would also be passive.

There were dozens of spots of arterial blood—the exclamation points with a head and tail that he believed could make that prediction through basic physics—on a man's coat and trousers, an outfit of David's that had been hanging on the inside of the bathroom door just inches from Allene's body. These were categorized as "impact or projected stains" because they typically traveled through the air and landed on an object.

"Measurements in a direct line from the point at which Mrs. Lamson's head was assumed to have lain show that all of these spots could have spurted from that point without interference," read details of his argument.

Oscar studied the trajectory of the blood from her head to the coat; the spurts were unobstructed and moving upward (the tails on the exclamation points were pointing up). This proved, he believed, that *no one* was standing behind Allene, because the killer's body would have created a void in the blood spatter on the coat hanging behind him in the tiny bathroom. There was no void—blood droplets were found up and down the coat.

Oscar addressed one of his most critical pieces of evidence, which was where Allene's body was located at the moment of impact. He examined the arterial blood spurts. One of the theories of bloodstain pattern analysis was that blood droplets traveled in a specific way. Oscar theorized that if Allene was hit from behind while standing in the tub, the tails of the blood spots would point downward on the suit hanging from the door. But the tails on these spots were all pointed up, as if they were released when she was hit at a lower location—like the porcelain sink basin. She might have slipped as she tried to climb out of the tub and hit the back of her head against the sink.

Oscar studied the pathologist's report detailing the injuries to the back of her head. Just one hit could have caused those four fractures, he believed, and if the impact of the basin had ruptured an artery, then it might explain such a large amount of blood loss. And microvessels in the scalp can bleed quite a lot when they're broken. The impact was great, and it wasn't clear to experts in 1933 whether a slender woman's body could produce that much force during a simple slip and fall. Even modern pathology experts can't say conclusively.

But Oscar was resolute. The forensic scientist checked over his notes again. His theory meant that there was no deadly weapon, no vicious murderer. David Lamson, Oscar told the defense attorney, was innocent.

Oscar settled in his lab once again to run more experiments. He analyzed the blood found on David Lamson's clothes. "Hemolyzed," he wrote. That meant that the fluid contained red blood cells that were broken open, perhaps when Allene's blood mixed with her bathwater. It seemed like a small point, but it would mean a win for the defense. There was no arterial blood on David's clothing. Allene's body was recounting the story of her death and helping Oscar untangle a mystery. An acquittal, Oscar assured himself.

Both sides agreed that there were several pieces of indisputable physical evidence. Allene Lamson had died in the bathtub that Tuesday morning on May 30, 1933. A head trauma caused her blood to gush profusely, splashing throughout the bathroom—the floor, the ceiling, and each wall. Some blood had mixed with water, while other spurts were undiluted because they came straight from her artery. And each medical expert agreed that just one fracture in the wrong place *could* create that amount of blood.

When police had arrived at 10:10 a.m., Allene's body temperature was warm and her bathwater had not yet cooled. Dr. Milton Saier gently shifted her head from side to side, checking for stiffening in her joints. Rigor mortis can begin as early as hours after the person died; it's still a key method for checking time of death. All muscles are immediately affected by rigor, but smaller muscles, like those in the neck and jaw, are affected first. Dr. Saier noted that Allene's neck joints were still loose. She had died within the hour, which jibed with David Lamson's story. Oscar thumbed through the autopsy report.

"Four lacerations on the back of the head, covering the occipital protuberance and surrounding it," it read.

The occipital protuberance is one of the bones of the skull, located at the back of the head. It protects the medulla, which controls functions like breathing and heart rate, and the cerebellum, which coordinates motor function and vision. If a blow crushes the occipital bone, it can cause a fatal brain injury.

"Four lacerations," Oscar read.

Lacerations, he knew, didn't necessarily mean gashes or cuts, but abrasions.

"Three of said lacerations were somewhat horizontal in direction, two, however, being somewhat curved," read the report. "One depressed fracture of the skull as well as an un-depressed stellate fracture."

One hard hit had likely killed Allene Lamson, according to her autopsy. Several smaller fractures resulted from that head trauma, like piercing the top of an egg with a pin and watching smaller cracks emerge. Three of the four lacerations were horizontal and parallel, while two were also somewhat curved. Oscar looked at the photos from the Lamson cottage. Pictures were visceral evidence for a jury; he had learned that from Martin Colwell's trial. They would illustrate his theory perfectly. He squinted at the bright white sink basin. The picture showed four irregular surfaces, each with a curved edge: the outer edge of the sink, a ridge, the edge on the inner side of the ridge, and, finally, the inner rim of the sink itself. The four edges of the basin matched the four lacerations on Allene's head. He called the defense attorney.

When the prosecutor learned Oscar's theory, he was infuriated. He argued that if Allene *had* fallen and hit her head on the basin, that much of her body would have remained outside of the tub. And how could she have flipped around and fallen facedown? David said he found her crumpled over the side of the tub before he cradled her, but he couldn't remember *exactly* where she was lying.

Oscar believed that her body's center of gravity propelled Allene back into the tub. Perhaps she tried to turn herself over after hitting her head? Or maybe she tried to brace herself on the sink? It was impossible to know.

The scientist fingered the strings wrapped around the dark skull in his hand. It seemed like a simple case: if there was any possibility that Allene Lamson might have died after hitting her head on the bathroom sink, then David was innocent. Even if both theories were plausible—murder or accident—which seemed more reasonable? Would a murderous husband really be capable of chatting up neighbors just minutes after killing his wife? Why would he invite a real estate agent and her clients inside? Common sense, Oscar Heinrich believed, would prevail in the jury room. But then again, murder was illogical—that's why motive was less important than hard evidence. And yet Oscar admitted that he might be employed by a wife killer.

He dropped his pencil on his steno pad. He was ready to reveal charts, photographs, measurements, and most important, he would disclose the truth. He wondered if his inventory list would be another obstacle like it had been in other trials. To most people, science was dull—horribly dull. Even newspaper writers, always armed with cheeky adjectives meant to arouse readers, seemed to struggle to make forensics sound provocative.

"Battle of Scientists Centers Around Dave," declared one paper.

"Heinrich Applies Benzidine Test on Floor," read another stale headline.

Even the picture's caption was tedious.

"Heinrich, a noted criminologist of Berkeley, California, made an investigation of the Lamson bungalow," it read, "and through a series of elaborately drawn charts will endeavor to trace bloodstain clues."

"Elaborately drawn charts" rarely mesmerized juries, Oscar admitted. And his public image wasn't buoyed by the media's contradictory descriptions. He was called a "nationally known criminologist and ace witness" in one sentence but "pale and deliberate" in another.

For a man who teemed with bravado during trials, Oscar Heinrich was still insecure, even after almost thirty years of experience. As he glanced over his photographs he worried that his stiff countenance on the stand could still be a hinderance once again, like it had been during the trials of Fatty Arbuckle and Martin Colwell.

John Boynton Kaiser worried about his friend's low morale. The librarian had spent years shoring up Oscar's confidence with much-needed reference books and indispensable advice tailored to craft a better version of "E. O. Heinrich" for court. Kaiser gently hinted at Oscar's pompous language during trials—and he inadvertently shamed an already-sensitive scientist.

"For years I have been enchained by the profound exactness of expression of my scientific mentors," Oscar complained to Kaiser, "can it be this that constitutes what you fraternally call my 'stilted style'?"

He *must* remain factually sound in the witness chair, Oscar argued to Kaiser; the panel deserved a frank, measured response to every question. And he was sure that jurors were brighter than attorneys believed. The jury would understand, even appreciate, his precise testimony in this case.

"I am both puzzled and concerned," he wrote Kaiser. "Do I persuade all shades of jurymen by being pompous, inflated, formal, stiff and bombastic? I cannot believe it, because they are too canny."

Kaiser warned Oscar, fragile from years of abuse on the stand, that he might undermine his own testimony by sounding too "scientific."

"Somewhere, somehow you have sensed an obstacle," Oscar wrote. "I beg of you to ponder to define it. If you say: 'be natural' I shall refuse to change."

But if he *did* refuse to change, Kaiser knew, then David Lamson might just be doomed.

The jurors jerked when Santa Clara County's district attorney smacked the weapon on the thick oak railing of their box. The metal resonated just inches away, four times. It was one of many satisfying, dramatic moments in the murder trial of David Lamson.

Oscar noticed the jurors as they watched Allan Lindsay palm that pipe during his closing argument—he was wielding the state's most important piece of evidence. Police found it lying in David Lamson's backyard bonfire, an item that might convict the quiet academic of murder. Beneath the smoldering pile of trash, investigators had uncovered a ten-inch piece of iron pipe—the perfect murder weapon, declared the prosecutor. He whacked the railing again. It seemed heavy, about three-quarters of an inch thick. The state had tremendous faith in that pipe.

The district attorney's case, which started on Monday, August 21, unveiled a litany of motives. But the most titillating storyline portrayed David Lamson as a cheating husband, a philanderer who carried on an affair with his daughter's nursemaid and made her pregnant. But that seemed unlikely, because Mary Dolores Roberts, still pregnant, actually appeared at the trial to help watch over Bebe. And she was now newly married to the baby's father. David would later lament about how the media had treated Roberts.

"I know nothing of her private affairs and consider them none of my business," he told reporters. "My only interest is profound sorrow that my own misfortunes should drag other lives into painful publicity."

Eventually, the state exchanged that motive for a lewder one. David, according to the prosecutor, was enamored of a beautiful and bright writer in Sacramento who had also attended Stanford. A decade after graduation, Sara Kelley and David collaborated on a series of books about the Great Depression for the university's academic press. Their working relationship required numerous meetings in Sacramento, which included dinners and meetings inside Sara Kelley's apartment (though her roommate testified that David never spent the night). He didn't appear to have hidden those trips from Allene, because she had noted one of them in her diary just months before she died.

David sent Kelley flowers several times, he said, because she needed to be photographed with them for her gardening column in the *Sacramento Union*. These were all neat explanations for what might have been a serious affair, Oscar knew, yet the motive was inconsequential because the physical evidence was the only thing that mattered. But the prosecutor knew that the motive *would* matter to the jury, particularly the women, so he designed his speech to sway the five female jurors.

"Everyone knows that if the husband starts to go out with a lady friend, it's not very long until the whisper comes home," said Assistant Prosecutor John P. Fitzgerald. "I don't believe there is a woman, a lady, present that is not going to be just a little bit sore, a little bit provoked, upon hearing of these conditions."

Untrue, David Lamson told his attorneys. They agreed and never called Sara Kelley to the stand, allowing jurors to speculate about their unusual relationship.

There were other theories from the district attorney; he claimed that David and Allene were close to divorcing and bickering over custody of Bebe. A deputy sheriff believed that he overheard David say to his sister at the cottage: "My God! Why did I ever marry her!" The claim was later refuted. David appeared to have small scratches on his face and neck, a sure sign of a struggle, but the defense claimed they likely came from trimming rosebushes.

District Attorney Allan Lindsay believed that these incidents all culminated with a violent argument on the morning of Allene's death. David demanded sex, and Allene refused because she was menstruating. When he suspected she was lying, he snatched a piece of pipe and hit her on the back of her head four times as she stood in the bath—a rage killing. He tossed the pipe in the bonfire to hide the evidence. And then the prosecutor also offered jurors one more example of his anger. David Lamson had killed before.

Almost twenty years earlier, thirteen-year-old David went on a hunting trip with another boy, Dick Sharpe, near David's family farm in Alberta, Canada.

"Sharpe and Lamson were shooting crows one Sunday afternoon when Sharpe stepped in front of Lamson just as he shot at a bird," read one newspaper report. "Soon after the accident young Lamson, with his mother and sister, moved to California."

Police in 1914 never charged the teenager because it was ruled an accident, but David Lamson, the prosecutor said, was capable of murder. And then Allene's leather-bound diary was submitted as evidence—a mortifying thought for anyone who had ever jotted down private thoughts in a journal. Now Allene's most intimate reflections were printed in newspapers around the world.

"Mother's Day with silk hose, candy, flowers and all," she wrote less than two weeks before she died.

According to the short passages of her diary, Allene was content with her husband.

"Dave packed and it was a beautiful trip," she wrote four months before her death, after a trip to the seaside with Bebe. "She wore Dave out climbing cliffs."

And then her private life became even more exposed. Just days after she died, the media unearthed some salacious gossip about Allene's love interests during college, other men who had been enamored of her. There was an engagement to another student during her first year at Stanford before she met David.

"Suddenly the betrothal was broken," reported one paper. "When Miss Thorpe returned to school here the following term, Lamson proposed."

Another story revealed Allene's sexual harassment accusations against a former janitor at the Stanford University Press who sent her endless love notes. Police considered neither man a suspect, but the tabloid fodder would have embarrassed any wife, particularly a private woman like Allene.

In early September, the Lamsons' friends, who were some of the most affluent in Palo Alto, filed past David. The couple had a loving, supportive relationship, they all testified. Very few would say a cross word about David Lamson.

During the three-week trial, both sides had batted the air quite a lot with the fire-blackened iron pipe. Over the summer, Dr. Proescher had analyzed the would-be weapon, claiming he had discovered blood in its tiny threads. Proescher had handed over a piece of the pipe to Oscar two months earlier, who braced himself for the result of his benzidine test. If blood was on the pipe, then his client was likely guilty. Oscar would have to resign from the case.

"The benzidine test was negative," Oscar concluded. "That settles the proposition that there is no blood."

Oscar's tests proved that there was an organic material on the pipe, but it was not likely blood—probably plant material or even rust. Other tests all had mixed results, so the substance on the pipe was inconclusive. As Oscar eyed Dr. Proescher in court, he felt uneasy. He noticed a disturbing pattern with the pathologist, a clear motive to manipulate evidence. When they tested clues together, the results were negative for blood. But when Dr. Proescher analyzed evidence on his own, the results were largely positive. The two experts were constantly contradicting each other. Which scientist would jurors believe? At the end of

August, both experts were called to testify, and neither man was assigned a flattering portrait in the media.

"Heinrich is meticulous of speech and rigorously impersonal and logical in manner," read one paper. "Proescher is excitable, leans forward in his chair, smiles and scowls with rapidly changing expression, and talks two languages at once at express train speed."

While Proescher's sketchy diction irked the overworked court reporter, it had amused Oscar. A bewildered newspaper writer turned to him and remarked, "You are fortunate—you speak German." Oscar nodded and smirked.

"I understand English and I understand German," he replied, "but I get along better when the two languages are not scrambled together like eggs."

Oscar glanced at the jury panel as they stared ahead. The members seemed dazed over phrases like "benzidine" and "hematoporphyrin." Oscar was antagonistic during the pathologist's testimony, often furiously jotting down notes for the defense team. He would glare at Dr. Proescher and then lean forward to quietly coach Edwin Rea, David's lead defense attorney. The portly man with small round glasses would then rise and relentlessly badger the doctor. Finally, the pathologist erupted when his testing methods were challenged.

"Do you doubt my integrity?" demanded Dr. Proescher, in his heavy German accent. "Do you think I'd put something else in there?"

"Well, now, doctor, that's a serious charge," said Rea.

He paused dramatically and then replied, "Well, yes, doctor, I do."

"Look out," yelled a spectator. "Somebody's going to be busted on the nose."

Dr. Proescher was clearly livid, but he remained quiet and professional—only his tapping foot gave him away. The confrontation had just compounded the anxiety in the courtroom.

The prosecution presented its most compelling evidence for the state's finale in early September—proof of murder. The jury eyed the long piece of pipe.

"One knock would crush the skull," Dr. Proescher testified.

Two other state experts said the same thing—one said the pipe was heavy enough to kill, "even if the weapon was wielded by a 12- or 14-year-old child." But another doctor, a prosecution expert who had performed Allene's autopsy, also admitted that an accidental fall was possible. No one could say for sure, and the state rested after almost two weeks of testimony.

When it was time for the defense to unveil its strategy, the attorneys made a controversial decision—David Lamson would testify. They believed he might connect with the jury by telling his own story. How could they believe such a bright, charming man was a killer?

But he would also be vulnerable to a shrewd prosecutor's cross-examination. David stepped onto the stand on Wednesday, September 6, 1933, and it was a mess. He was a dreadful witness for his defense attorneys but a godsend for the district attorney. David's answers were vague—his memory was hazy.

"What was your wife's position in the tub when you first saw her?" asked Allan Lindsay.

"She was half in and half out of the tub," David replied.

"Just where on the tub?" Lindsay asked.

"Towards the middle of the tub, I believe. I don't have a clear recollection."

"Why didn't you call a doctor?" asked Lindsay.

"I don't know how to answer that, Mr. Lindsay," he meekly replied.

David repeated, "I can't remember" and "I don't know how to answer that" more than a dozen times. As he slid into the chair next to his attorneys, the prosecutor reminded the jury that he had been an amateur theatrical actor. The case would rest on Oscar Heinrich's expert testimony. The forensic scientist gripped his charts and photos as he sat near the defense attorney.

The clash of the experts reached a crescendo by Saturday, September 9, when the defense began calling its own experts. Oscar smiled at a familiar face—he was relieved to have an ally, a well-known and respected physician who supported his belief that Allene had died from an accidental fall. Dr. R. Stanley Kneeshaw was the first defense witness to defend Oscar's accident theory. The fractures, he testified, were not unusual for someone who had fallen.

"The injuries are typical of what we call an explosion fracture," Kneeshaw testified, "such as results from contact with a flat surface."

"Do you say it is impossible that those wounds could not be produced by that pipe in the hands of a man weighing 180 pounds?" retorted Herbert Bridges, for the state.

"I would say it would be extremely unlikely," replied Dr. Kneeshaw.

Bridges sneered.

"Would you let me try it over the back of your head?" he quipped.

"No, but I'd make a wager with you if you wanted to try it on someone else," Kneeshaw calmly replied.

And that was the theme for much of David Lamson's murder trial—a zealous prosecutor paraded dozens of witnesses and experts before jurors, while a disorganized defense team struggled to deflect ungrounded rumors. Science became a minor player.

Oscar Heinrich settled into the low wooden chair to the right of the jury on September 9. The gold chain of his pocket watch dangled near papers on his lap. Edwin Rea asked him to refute Dr. Proescher's claims that blood was present on the iron pipe found in the bonfire. Oscar was pleased to refute just about *everything* that the pathologist had claimed.

"When blood is charred completely, there is no test for absolute verification that it is blood," said Oscar. "The three most delicate tests cannot be used at all. Others leave too much room for error from other elements, which may give the same reaction as blood and may be in the ashes."

He was persuasive, noted reporters. Jurors seemed engaged as he prepared to unveil his most convincing evidence—his series of large photographs. Oscar was pleased with how they turned out. The jurors in the Martin Colwell murder case had swayed him about their importance.

Oscar pointed to the pictures from the reenactments inside the Lamson cottage starring his assistant George Weber's wife. He sat in the witness stand in a smart dark striped suit, legs crossed.

"Doctor, did you perform an experiment at the Lamson cottage demonstrating a fall?" asked Rea.

"I did," answered Oscar.

He began to swivel to his left to address the panel, but he was silenced immediately.

"The only question is how well we know whether we know it is a similar experiment," said Judge Robert Syer.

Oscar glanced at the defense attorney—he knew to stay quiet.

"The conditions are the same in that bathroom," insisted Rea, "that we want to show this jury is you can take a woman the same height as Mrs. Lamson."

"Objection!" yelled the prosecutor.

"Your Honor, it is something very important in the case, that goes to the very gist of the case," pleaded Rea.

"The only way to prove it would be to kill someone," insisted Assistant Prosecutor John P. Fitzgerald.

And with that argument, the judge barred Oscar from testifying about his experiments in the bathroom using a live model. It was disastrous to David Lamson's defense and ridiculous, thought the criminalist. The academic at the defendant's table sank into his chair while his attorneys hoped for more luck.

Oscar brooded quietly on the stand—the ruling was an affront to his hard work. He was incredulous that anyone, particularly a judge without a scientific background, would distrust him. But Oscar remained impassive to the jury, knowing he could offer other evidence, even more reliable proof.

He carried a long wooden pointer toward the blackboard, where he drew blood spots with little tails facing different ways. He spent hours detailing how the shape and consistency of blood spots could verify a cause of death. Courtroom gossips whispered about his gruesome

Holmesian experiments in the seclusion of his lab—it was rumored that Oscar had sliced an artery in his own arm for research. He slyly denied it, but admitted to orchestrating other bloodstain pattern experiments using veins from other live volunteers.

Oscar pointed to a large photo of the inner door, an exhibit that showed ten blood spots. Jurors nodded, and Oscar's confidence seemed restored. But when he tried to describe where Allene's body lay, Judge Syer stopped him once again. Oscar seemed confused. The judge said that Oscar's personal opinion was not relevant because he had visited the scene two weeks *after* Allene's death. The majority of the judge's rulings were against David Lamson—Oscar felt hopeless.

Exhausted, Edwin Rea rested his case. Now David could only hope that the jury believed Oscar Heinrich, just enough to create reasonable doubt. On Saturday, September 16, the prosecutor issued a forceful edict to the jury during his closing arguments.

"We are not going to let this man walk off to go out of here with blood upon his hands," said Allan Lindsay. "Help us stop this wave of crime. If you don't, it simply goes out to the world a man can commit a crime and get away with it. More women, more children will suffer."

Lindsay cradled the iron pipe and raised his voice.

"Why, you couldn't fracture a head?" exclaimed Lindsay. He slammed the pipe against the railing.

"I am not using my full strength," he said. "Put that in the hands of a one hundred and eighty pound man."

The prosecutor struck the railing three more times and stared at the jury.

"How would you like to have me hit each one of you?"

Nelle Clemence was anxious as she listened to the foreman review the evidence. There were papers on the wooden table mixed with the police's crime scene photos. Notepads covered with scribblings lay to the side.

The white-haired, middle-aged juror was a retired fruit rancher, a businesswoman with strong convictions and a wicked sense of humor, though the local newspapers assigned to her a less appealing description.

"One can imagine that she will make arrangements to have the cat fed while she is locked up on the jury," reported the *Oakland Tribune*.

Clemence was normally bold, but just a few hours into deliberations, she was suddenly hesitant. Jury rooms in the United States could become uncomfortable, particularly during capital murder trials. Twelve strangers with likely little in common were isolated for days or even weeks, compelled to discuss the details of a disturbing death. Since Colonial America, jury rooms have hosted passionate arguments, personal jabs, and even physical confrontations all later blamed on the hubris of men and women divided over the rightful blame for a horrible tragedy. Unfortunately for David Lamson, he was not blessed with that sort of introspective jury. And that was also unfortunate for Nelle Clemence.

On Saturday, September 16, 1933, the jury of twelve retired to the chamber inside the Santa Clara County courtroom to deliberate. After several hours of discussion, the panel adjourned for lunch, but right before they left, the foreman, George Peterson, requested an oral ballot. Clemence squirmed—she was the only "not guilty" vote. She looked at each of them. They were all so certain that David Lamson was a killer. The foreman glanced at the older woman through his round glasses. A merchant from Saratoga with a dark toothbrush mustache, Peterson seemed almost zealous in the jury box. Courtroom observers noticed how attentive the stout, older man seemed as the prosecutor laid out his case.

"My mind was made up almost a week before the case was submitted," Peterson proudly told the press later.

He seemed to have no idea how outlandish that statement must have sounded to anyone who believed in the integrity of the American judicial system, but the other jurors agreed with him.

Even though the judge had banned access to all media, two jurors (including Peterson) flipped through newspapers during a picnic at Oak Dell Park six days into the trial. The jury had been sequestered, but jurors went to movies, attended public lectures, and even visited with their spouses at the jurors' hotel, the Sainte Claire. The youngest woman on the jury, alternate Ora Conover, went to a party during the trial and declared, "The defendant was guilty and should be hanged without a trial."

George Peterson, the passionate foreman, lied during jury selection when asked if he was friendly with the sheriff in charge of Lamson's investigation. "No," had been his response. But Peterson had been a guest of the Santa Clara County sheriff during a private tour of San Quentin State Prison—a clear conflict of interest.

As the jury discussed the evidence over lunch, Foreman Peterson seemed alarmed over talk of a "not guilty" vote. He issued a threat to another panelist across the table from Nelle Clemence.

"You know what happened to the juror that was fixed in the Matlock case," he warned.

Just two months earlier Joe Matlock had been convicted of gunning down a police officer in San Jose, California, during a traffic stop. Eleven jurors voted for the death penalty, but there was one holdout. The fifty-seven-year-old dissenter argued that Matlock deserved life in prison, not the noose. The debate turned vicious as vote after vote was cast and James Howard refused to waver. After twenty-nine ballots, the other eleven jurors abandoned the argument and agreed to a life sentence.

Within hours, Howard's life story was detailed in newspapers across the country—he was heavily criticized and even threatened. And now two of Matlock's jurors were serving on David Lamson's jury. Nelle Clemence felt powerless as she sat quietly in her seat during lunch. After the meal, the other eleven jurors seemed exasperated. They all argued with Clemence and, as a group, "attempted to convince her of the guilt of the defendant." They explained that David killed his wife in a rage after being denied sex the night before.

"Impossible," insisted Clemence.

She glared at the male jurors and wondered if they would bludgeon their own wives if they were denied sex. The men were enraged.

"You know David Lamson is a moral degenerate," hissed J. A. Harliss.

Peterson and another member yelled that they would vote "guilty" to "protect their daughters and society from such fiends." They accused Clemence of shunning the safety of the public while two jurors furiously scribbled notes, threatening to keep them as evidence.

"She can be held for contempt of court," declared Viola Brown. "We can send for one of the alternate jurors and have her disbarred entirely. We had better have her replaced and have another juror sent in. That is what alternate jurors are for."

Of course, she was wrong. And while most of the arguments were ludicrous, their attacks against Clemence had been relentless. Juryman George Hegerich assured her that it was actually *illegal* for the district attorney to prosecute an innocent man for murder.

To further convince Clemence to change her vote, Foreman Peterson settled on circumventing the court's strict rules. The jurors built a crude model bathroom in the jury room with chairs and desks, all meant to replicate the Lamsons' own tiny bathroom. There were a few issues, most notably the absence of walls. The real bathroom was very small, with so little space that it seemed impossible for anyone to stand behind Allene and swing a pipe with much force. Nonetheless the prosecutor's maps, charts, and crime scene photos were rounded up and delivered to the room. One by one, the twelve jurors "fell" from the "bathtub," each trying to simulate Allene Lamson's accident.

"All of us tried to fall out of that tub in some way so that we could hit our heads on a washbowl hard enough to dash our brains out," said Hegerich. "We succeeded only in convincing ourselves that it couldn't be done."

In contemporary times, that sort of exercise in a jury room would likely cause an immediate mistrial. David's defense attorneys called it a careless, dangerous experiment—unsupervised and uneducated. The judge had refused Oscar's experiment, yet the jury seemed free to conduct its own.

The jury also discussed the defense's star witness, the most crucial expert in the trial. They debated Oscar's bloodstain pattern evidence, his complex arithmetic that required a cardboard dial and some string—the proof that David Lamson was not a killer. And they rejected all of it.

"We knew that dead women don't jump around, spattering blood over everything," said Hegerich. "E. O. Heinrich, the defense criminologist, didn't convince us of anything."

Oscar's entire testimony was unsound and even ridiculous, according to jurors. They each gripped the iron pipe found in David's bonfire, studied its weight and length.

"We looked at it, practiced hitting with it," said Hegerich. "Then we remembered that Mr. Heinrich had said that those blows couldn't have been inflicted with that pipe and we knew he was wrong there."

The jurors had misunderstood much of the expert testimony from both sides. The argument was not that a pipe *couldn't* kill someone—it was that a pipe had not likely caused *these* fractures. The key for jurors was not David's criminal history, his arguments with Allene, his friendship with Sara Kelley, or his erratic memory of details with police or family members. The motive didn't seem to influence their discussion, and that was, in fact, the sound path to follow. The cornerstone of the jury's decision rested with their understanding of the physical evidence. And that was dire news for David Lamson.

"Lamson virtually convicted himself," insisted Peterson. "If he had not washed her body and then put it back in the tub he would not have been convicted."

The jury had remembered the phrase "washed blood," a description Dr. Proescher used many times during the trial. Jurors thought it meant that David had *washed* Allene's body, but "washed blood" actually referenced how bathwater had diluted Allene's blood. The jury misunderstood the meaning, and so had the press.

"There was evidence of attempts to wash out the blood," reported the *San Jose Evening News.* "This was taken to imply that someone had tried hurriedly to hide evidence of a crime."

After arguing all day, the jury was escorted by bailiffs to dinner, and as they walked back, a female deputy sheriff invited Clemence to walk ahead with her.

"I know David Lamson is guilty," whispered Leonora Ghetti. "I know it is to be a fact that David Lamson is a moral degenerate. I also know that he confessed to his lawyers right from the beginning, but the lawyers would not allow him to go on with the confession."

Ghetti had continued to harass Clemence by plying her with false information about David's sister, the physician.

"Dr. Lamson was bringing in dope to Dave ever since he's been in here," insisted Ghetti. "Dr. Lamson says that her brother is a hopeless case, and that she has had a lot of trouble with him."

When they reached the courthouse, the bailiff opened the giant wooden doors and the jurors filed through. The lock turned behind Nelle Clemence. The others eyed her as the foreman demanded a third vote—a final ballot. Clemence scribbled her answer, and it was now unanimous. The jury declared David Lamson guilty of first-degree murder. Nelle Clemence had to relent.

The jury could have offered leniency by recommending life in prison instead of the death penalty, but the panel chose to send David Lamson to California's Death Row. His fate had rested with suspicious jurors who were confounded by a scientific expert's strange bloodstain pattern experiments. Oscar Heinrich's testimony had failed David Lamson, and now he was destined to die in the gallows. The criminalist was disgusted, then distressed. And he blamed Dr. Proescher, yet another rival who had undermined him in court.

"We are in for a critical winter which may even pinch us and cause new means to be sought to keep ahead of the sheriff," Oscar Heinrich wrote his son Theodore in 1933. "I have been growing more and more disgruntled day by day over the failure to receive your financial report."

The stress of the David Lamson case had tested Oscar's patience, particularly as he was trying to encourage his son to be responsible with his limited budget. Theo was attending graduate school at the University of Cambridge in England, one of the most expensive colleges in the world, while Mortimer was now an undergraduate student at the University of Oregon in Eugene. Despite the cash windfall from the Lamson case, Oscar's business had been reduced by more than half. He was beginning to resent his twenty-three-year-old son's frivolous spending.

"The steady drain on our resources to maintain your program has been and is most devastating to our economic security and to our hopes for Mortimer's program," he wrote Theo. "At first glance it would appear that you were over-indulgent of yourself with your funds."

Oscar's grip on his sons' lives was tightening as they grew older. He began criticizing Theo's expensive European trips, his travel companions, his curiosity about varied careers, and even his choice of women. "A good wife during such a period can be, like your Mother was to me, a gorgeous inspiration," Oscar told Theo. "But platinum blondes, born with golden feeding spoons, can't qualify."

The criminalist seemed to be convinced that his sons, without his guidance, would meander through life without embracing responsibility or earning professional respect.

"Please do not allow a woman to rush you at this critical stage in your career," he told Theo. "Get your academic business done and tell the girl that for the moment you have work to be finished."

Tiny shards of glass lay on the concrete. The older man delivered a superb upper cut to his opponent's jaw. They both bellowed profanities, wildly flinging their bodies as a crowd of more than one hundred gawkers retreated to give them sparring room. Both men had deep slices on their faces. The scene was appalling—and also quite embarrassing, because the two boorish men were middle-aged, highly educated attorneys who were fistfighting in the Santa Clara County Courthouse . . . all over a woman.

"I'm sixty years old, but I'll take you on right now," screamed defense attorney Edwin Rea as he flailed on the ground, the victim of a knockdown shot.

"You ought to be disbarred!" bellowed the county's prosecutor Allan Lindsay before Rea sprung to his feet and lunged.

At forty-five years old, Lindsay was fitter, but Rea boasted of holding a heavyweight championship boxing title while at Harvard University. After exchanging about a dozen blows, the only person who could break them up was Allene Lamson's brother, and he still needed help from some scrappy court officials and a few brave newspapermen. One week after a jury had convicted David Lamson of first-degree murder, both sides were now back in court thanks to Nelle Clemence. She was the lone juror who believed in David Lamson's innocence—the one who voted "guilty" after nine hours of pressure. In a sensational affidavit "bristling with charges of coercion and misconduct," Clemence accused the other jurors of bullying her during deliberations.

"I was always convinced of Lamson's innocence," said Clemence in her affidavit.

Deputy District Attorney Lindsay was incensed and accused the defense attorneys of conjuring up a false report based on hearsay.

"Are you calling me a perjurer?" yelled Edwin Rea.

Minutes later the "battling barristers," as the newspapers called them, were swinging at each other while David Lamson was led back to jail. Edwin Rea accused prosecutors of misconduct and blamed the trial judge for numerous errors, like barring Oscar Heinrich's accidental fall experiments. The defense attorneys also charged the judge with giving "prejudicial instructions" to the jury. Rea demanded a new trial, calling the conviction "the most shocking, the most hideous verdict in the history of Santa Clara County."

When David arrived to court on Monday, September 25, he was upbeat. He chatted with his worried attorney and said, "Don't worry, everything will be all right," but as the arguments for a new trial began, David's demeanor changed. Superior Judge Robert Syer, the same judge who had sat on the bench for his trial, rejected each of the defense team's accusations.

"Lamson's once straight shoulders bowed and the smile became a frown," reported one newspaper.

The next day Judge Syer denied the defense's motion for a new trial and immediately pronounced David Lamson's death sentence.

"On Friday, the 15th of December, 1933, you [will] be hanged by the neck until you are dead," said Syer. "May God have mercy on your soul."

David gently bowed toward the judge.

"I know in my heart that I have been a good husband to her," he told Judge Syer. "I have done her no harm, I am innocent of her death as you yourself. That is all."

It was an eloquent and composed remark from a man whose calm disposition had unnerved members of his jury. They spent weeks leaning forward in their seats, hoping to glean some emotion from a would-be killer, but they learned little.

"My boy doesn't wear his heart on his sleeve," insisted his elderly mother, Jennie Lamson. "The public has horribly misunderstood him because of his temperament, one of which I am proud as a mother, whatever his fate may be. He can't bear to make an appeal to sympathy."

Judge Syer seemed unmoved by David's declaration of innocence, and he was soon led away by the bailiff, followed quickly by his tearful mother and sisters.

"I did not do it," David insisted.

"They can't hang my boy," his mother cried to reporters. "You boys know that. They can't hang my boy."

For four months, David's sisters faithfully sat on the hard wooden benches behind him and pursed their lips as the prosecutor made horrible accusations. They refused to believe he would kill his wife, and now they were heartbroken because they might lose Bebe, too. Allene Lamson's brother, Frank Thorpe, had petitioned to adopt the little girl. She was living in Palo Alto with Dr. Margaret Lamson, David's sister, but Thorpe and his family wanted to change Bebe's name and relocate her to Allene's hometown of Lamar, Missouri, almost two thousand miles away.

"Allene [when she was alive] asked me to take care of little Allene Genevieve rather than her own mother," said David's mother, "and that is what I am going to do as long as I am able. To me that baby is a sacred trust."

David's attorneys requested that the court enter a motion to appeal the trial and its verdict to the Supreme Court of California, a panel of seven judges in San Francisco that would consider if the trial was adjudicated fairly. Oscar Heinrich felt rankled and bitter.

"The verdict means nothing to me, except that I regard it as a terrible miscarriage of justice," he told reporters.

August Vollmer, Berkeley's former police chief, was outraged. In the press, he praised Oscar's bathroom experiments, vehemently defending his closest ally in a lab coat.

"Any intelligent person who would visit the Lamson bathroom and see the demonstration put on by E. O. Heinrich would be convinced that the death was accidental," he declared.

Oscar felt sickened by the jury's conclusion, as well as by the judge's baffling decisions.

"I took the stand to explain as an expert how it was impossible to establish Lamson's guilt from those facts, which clearly proved his innocence," Oscar said. "But objections by the prosecution prevented me from submitting my most valuable deductions."

Oscar had no time to grieve for David Lamson because there were other cases to solve. He lamented to Vollmer, the smartest cop he knew.

"Instead of providing an opportunity for presenting the truth," Vollmer said, "it is a game for lawyers."

Vollmer believed in justice—and he also believed in Oscar Heinrich's sound evidence. There's something we can do, Vollmer promised.

His thick, wavy brown hair was gone, now lying in bits on the floor of the "fish tank," the intake room for new inmates. It was Friday, October 6, 1933, when thirty-one-year-old David Lamson began his life on California's Death Row. The former executive at a prestigious university press became Convict No. 54761 in San Quentin Prison.

David exchanged his smart brown tweed suit for a dark gray prison uniform. Guards ordered him to stand against a wall by an angled mirror as a photographer snapped his mug shot. The metal board in front of him declared the day of his scheduled execution: December 15, 1933, just two months away. His fingers were black from ink. He was frightened or furious—or both. Mostly, he just seemed subdued. The press, which was always attentive at his trial, continued to report on his transition to prison. "Stoic," noted some reporters, but also resolute.

"I am innocent," he told the press. "I am sure exoneration must come."

David joined seven other men on the Row, all awaiting capital punishment. His cell, No. 22, had recently hosted a convicted killer who proclaimed his innocence and spent several years dragging out his death sentence until the courts denied his final appeal. Two weeks earlier, guards led the twenty-three-year-old man from No. 22 and executed him. This was not a lucky cell.

David placed a photo of Allene and his daughter on his table. He examined his wife's wedding ring, one of the few personal items he was allowed to bring inside. His mother and sisters continued to fight with Frank Thorpe over the custody of Bebe—the families agreed to pause the court battles until after David had exhausted his appeals. His mother was weary from the emotional angst of losing her son.

But there was one bit of optimistic news—the list of prominent public supporters for the convicted killer had grown longer, and the

Lamson Defense Committee was created, led by August Vollmer. One of the nation's most respected cops was determined to prove David's innocence. The committee quickly expanded to dozens of members, including professors, writers, journalists, physicians, and clergy. Vollmer collected donations for more research. Oscar remained quietly in the background—it would be improper for an impartial scientist to join such a vocal group. But he encouraged Vollmer to not give up. The members published a persuasive report called *The Case of David Lamson,* a 119-page indictment of the prosecution's case against an innocent man.

"It exemplifies the frightful potentialities of a complex machinery of law when set in motion by persons over-zealous, injudicious, and incompetent to control what they have started," read the report.

And then David Lamson's outlook shifted from cautious optimism to true hope. His defense team filed a six-hundred-page brief that, along with the committee's report, convinced judges with the California State Supreme Court to delay his execution until they could consider his controversial case. David Lamson's life now depended on the judgment of seven learned men in San Francisco.

It was a cool, sunny October day in 1934, Saturday the thirteenth, when David heard the news from his attorneys, one year and one week from the day he had entered San Quentin. He was dazed. California's State Supreme Court justices had made a wonderful finding in his favor.

"Every statement of the defendant, capable of verification, tends to support his claims," read the statement representing the principle decision. "Where in the case before us is there evidence of a preparation for commission of a crime? Where is the plan by which the death was to become known? Where is the effort to prevent detection or for concealment? Where are the bloody or missing clothes? Where is evidence of a seizure in order to inflict the blows?"

It was a circumstantial case, believed the panel, one that did not prove beyond a reasonable doubt that David Lamson had killed his wife.

"The macerated areas of these lacerations form a distinct pattern," they wrote. "How could an iron be wielded with such precision?"

A pattern had swayed their decision—not a bloodstain pattern, but a *wound* pattern. The judges overturned David's murder conviction and immediately ordered a new trial. It was a tremendous victory, but not quite vindication.

"It is true that he may be guilty, but the evidence thereof is no stronger than mere suspicion," they said. "It is better that a guilty man escapes than to condemn to death one who may be innocent."

The decision illustrated the contradiction of the justice system, then and now. The majority of the seven judges suspected David *was* guilty, but it was more important to preserve the integrity of the legal system than to convict a potential murderer. If the prosecution couldn't prove its case, then a guilty man *should* walk free—a difficult concept to accept. But regardless of his innocence or guilt, several of the judges concluded that Oscar's experiments would have helped offer David Lamson a fair trial.

"We brought into our chambers the wash basin and the pipe, as well as other exhibits and examined them," said Chief Justice William H. Waste. "Justices Shenk, Spence and I felt a serious error had been committed in not allowing experiments that might, in the jury's mind, have eliminated the question of Lamson's guilt."

The Supreme Court judges couldn't resist conducting their own uneducated experiments. Could pretending to fall actually replicate a deadly accident? Three judges believed it would, but the other four justices did not.

"The other justices did not concur in this view on the grounds that the circumstances could not be faithfully reproduced in an experiment because they argued you would have to kill the subject," said Waste.

And that was the central problem with those types of demonstrations in court, even when they were conducted by an expert like Oscar Heinrich. The state's highest court couldn't agree.

The Lamsons were elated that he would receive a new trial, no matter the reasoning behind the decision. While David's mother was euphoric, his three-year-old daughter had no idea what was happening. Jennie Lamson refused to tell Bebe that her mother was dead and that her father was condemned to die for her murder.

"She still believes that her mother has 'gone away for a long time,'" said one paper, "and that her father is 'in the hospital.'"

Allene Lamson's family in Missouri was still fighting for custody of the little girl, and when they learned of the court's decision in David's favor, they were infuriated.

"We have never changed our opinion of Lamson's guilt," wrote her brother and their parents. "We believe that the granting of a new trial will result only in a more positive conviction."

David's defense team requested that the district attorney drop the charges and release him, arguing that there was no new evidence. Superior Judge Robert Syer, David's original trial court judge, denied the request and ordered a new trial—and then denied David bail. He would have to return to the jail in San Jose, back to where he had started the fight for his life.

It was such an odd feeling—the thrill of returning to his old cell. But a month after the California State Supreme Court's decision, David Lamson walked out of San Quentin clean-shaven and wearing the same brown tweed suit he had worn on his first day about fourteen months earlier.

"I'm so happy," David told reporters, "I don't care what happens now. When I passed this way before, I thought it was to be a one-way trip. It is good to see the water again, from the deck of a boat. I don't mind the rain. It seems like a pretty good day to me."

Hours later he walked through the doors of the Santa Clara County Jail and into the arms of his two sisters and their mother. They had a special dinner, which concluded with pie a la mode in a private dining room, but Bebe was still missing. David was reluctant to see her—it would be too painful to leave again if he had to return to prison. Bebe had hoped that her father would return from the "hospital" in time for Christmas next month. At night, David sat in his cell near a borrowed typewriter. The keys clicked by lamplight.

The defense attorney shoved papers at the prosecutor, some insurance company statistics that showed that thirty thousand people a year died from injuries after bathtub falls. David's team hoped the prosecutor would decide to release him because of a lack of new evidence, but the district attorney refused. He was determined to send David Lamson to the gallows.

Jury selection began on February 18, 1935, and immediately after the trial began, the defense was delivered a gift, almost like divine intervention. One juror actually slipped and fell in the hotel bathtub, leaving bruises across his body. David's defense attorney quickly capitalized on the news.

"McKenzie immediately—and by no means silently—arranged for the hotel to install safety mats in the tubs of all the bathrooms of the jurors," said one newspaper report.

Oscar Heinrich smiled at the wonderful news from the judge. He was finally able to present his slip-and-fall evidence and bloodstain pattern analysis without interruption. And he could also build a replica bathroom in the courtroom to help explain his accident theory.

"In rolling off, she slid into a position hanging across the rim of the tub," Oscar explained, using his assistant. "We repeated the procedure three different times, and each resulted in exactly the same way. The position is very close to the same position which I had found was the focal point of the blood spray."

His testimony was convincing, but not convincing enough. The jury deadlocked 9–3 *for* conviction. David was devastated but hopeful, because perhaps now the district attorney would realize that a retrial would be pointless. But no—there would be a third trial. Oscar was fatigued, and for the sake of his health, this case had to end soon.

As David sat in his jail cell waiting for his November court date, he edited the rough draft of his manuscript. It was a 338-page nonfiction book about the condemned men of San Quentin, those convicts who waited for the noose alongside him for more than a year. Months earlier Charles Scribner's Sons, a major New York publisher, bought the rights and now newspapers across the country printed large excerpts containing tight prose crafted by a convicted killer.

"They say that when a high-velocity bullet strikes you, you feel no pain for a time—only numbness," David wrote about the day he learned his verdict. "This was like that. I could feel the muscles of my face stiffen. I don't believe I thought at all, beyond willing myself to keep my face and body still, no matter what. To sit still and take it."

In September 1935, as David awaited his third trial, his book was published. *We Who Are About to Die: Prison as Seen by a Condemned Man* became a bestseller, and public opinion seemed to sway his direction. Readers were fascinated by a narrative that lambasted a corrupt judicial system but also humanized the men on the Row.

"The frog at the bottom of the well thinks that there is only four feet of sky, because that's all he can see," David wrote. "I have neither the desire nor the ability to write a positive book. I stayed in Quentin too long for that."

Literary reviewers across the country praised his writing, especially his candor. The *Los Angeles Times* gushed over his vivid descriptions. Critics valued a fresh point of view that was offered in a uniquely eloquent way.

"It is really an appeal to consciousness not to mercy," read the *New York Times*. "It is a hard lance, aimed at our shields of ignorance, and it is to be hoped that it will break through."

Alexander Woollcott, a legendary book critic for *The New Yorker* magazine and a literary icon, called the book "profoundly important. It is the book to which my thoughts have gone back oftener than to any other book I've read this year."

Instead of traveling from city to city on a book tour, the most popular debut author in America spent weeks staring at the bars attached to his window. David Lamson had little time to enjoy his literary success.

In November 1935, a third trial was scheduled, but there was an error with jury selection, so the judge declared another mistrial. Undeterred, the prosecutor ordered a fourth trial, and by this time Superior Court Judge Robert Syer had recused himself. David had now spent more than a year in the jail in San Jose and more than two years behind bars. And he had still not seen his daughter, even though she was just a few miles away with his mother and sisters.

In March 1936, almost three years after Allene's death, a fourth trial was held, proceedings that repeated essentially the same evidence and the same witnesses from previous trials. The jury deadlocked once again, 9–3 for conviction. Jurors *still* didn't fully believe the new scientific techniques, but they also were not convinced by the prosecutor's case. Thirty jurors across four trials had believed that David Lamson was guilty—and the prosecutor considered that when he debated trying him for a fifth time. But in April 1936, after spending almost $70,000, the district attorney relented.

"With tireless effort we sought to gather legal evidence," said the prosecutor. "We are unable to produce any new evidence."

David collapsed in his jail cell's chair when he heard the news. After two and a half years behind bars, including thirteen months on California's Death Row and three full trials, it was over. On April 3, 1936, David Lamson was released.

The five-year-old stared for a moment at the handsome gentleman in the dark suit.

"Who is this man?" she asked, looking over at her aunt.

"That's your daddy," Dr. Margaret Lamson sweetly replied.

Bebe glanced at him and her eyes widened.

"Oh, Daddy," she cried, and ran into his arms. "Where are you going to sleep?"

David grinned. Her vocabulary was so advanced now, one of many big distinctions between a two-year-old and a five-year-old. He kissed her and listened to her litany of statements, questions, and demands.

"You need a shave, Daddy," she replied. "Please read to me right away."

And that was that. After three years of separation, David Lamson began to plan a new life with his daughter. After taking a short vacation David turned to the pages he wrote in his San Jose jail cell, the bestselling book that would cement his legacy after prison. In 1937, *We Who Are About to Die* was adapted into a movie with the same title. David, the perennial writer, helped with the screenplay about a man who was kidnapped by mobsters and then framed for the murders they committed. Condemned by a greedy prosecutor with political aspirations, the man endured Death Row while his friends worked to free him.

David Lamson remarried the year of his release, and his wife, another writer, adopted little Bebe. After the premiere of his movie, David authored another bestselling book, a novel called *Whirlpool,* which was a fictionalized account of his case. David became a professional writer, publishing more than eighty magazine stories in well-known magazines. In 1975, he died at his home in Los Altos, California, at age seventy-two.

He had earned tremendous professional success after his release, but David Lamson seemed to continue to carry along a weight. His daughter once spoke to a newspaper reporter about her father after he died.

"He never discussed the specifics of the case," she said, "but he always spoke lovingly of my mother. I had a very normal and happy family life growing up, but the experience clearly had a deep psychological impact on him for the rest of his life."

By the time David Lamson left San Quentin in 1934, he had grown fond of some of the inmates. But he was also disgusted with American courts. He was free but not exonerated—like a man forced to look at the scars left by his missing shackles. And he was consumed with guilt when he remembered the day he left San Quentin State Prison.

"I looked up at the Row. I waved back to them . . . a little shamefacedly because I was fortunate," he wrote in *We Who Are About to Die.* "You feel helpless and bitter, and the one feeds on the other until they merge into an angry, cursing impotence the worse for being throttled."

Oscar Heinrich was also deeply frustrated with America's justice system. Despite all his efforts, science was far from accepted in court. But there would be hope.

In 1935, the year between Lamson's first and second trials, a biochemistry professor at the University of California at Berkeley with an interest in forensic science contacted Oscar—Dr. Paul Kirk. He asked Oscar to present a paper for the American Chemical Society at its microchemical symposium, an event for chemists who specialized in testing minute amounts of materials, like evidence at a crime scene. Oscar confided in Dr. Kirk about his jury problems in the Lamson case and the struggle of all forensic experts to teach science in court.

"As you no doubt have observed in the Lamson case," Oscar wrote Kirk, "the micro-analyst has a difficult time to present something to a jury which they do not readily understand."

Oscar had learned a difficult lesson over his decades in forensic science. "What they do not easily understand they reject," he told Kirk.

Oscar suggested that biochemists, like Kirk, learn from his own mistakes and train chemists to simplify their testimony about complicated tests. Five months later, Kirk was hired for his first murder case—the killing of an elderly shoe cobbler in Tucson, Arizona. Police asked him to analyze hair found in the dead man's hand, but the case was never solved. With Oscar's recommendation, August Vollmer encouraged

Kirk to join his faculty as a criminalistics professor at UC Berkeley. By 1955, he was a forensics expert who would become a star witness in one of the most infamous trials in American history, the murder trial of Dr. Sam Sheppard in Ohio. It was the case said to have inspired the TV series and feature film *The Fugitive*.

The similarities of the cases of David Lamson and Sam Sheppard were startling: two prominent, handsome men, both married to beautiful women and both on trial for beating them to death. The national media chased each case, publishing salacious headlines accusing both men of being philanderers. Sam Sheppard claimed he struggled with a "bushy-haired" intruder who later escaped. The jury didn't believe the neurosurgeon and gave him the death penalty.

Sheppard spent ten years in prison before the United States Supreme Court ordered a new trial, contending that the press coverage had created a "circus-like atmosphere." In the 1966 trial, Dr. Kirk testified that the bloodstain patterns from his wife's wounds had suggested that Sam Sheppard was not the killer; Kirk analyzed the blood found on the doctor's wristwatch, the sprays on the walls, and the height of the various spots. All this evidence, Kirk believed, pointed to another killer. The jury believed him, and Sheppard was acquitted.

During the case of David Lamson, Oscar Heinrich faced formidable experts in court, pathologists with expertise at crime scenes and inside autopsy rooms. But in our modern judicial system, their medical credentials should not be enough to qualify them to testify in most criminal cases about bloodstain pattern analysis. BPA demands expertise in applied mathematics and physics, among other complicated scientific principles. Pathologists don't typically have that type of training, but Oscar Heinrich did. With his chemistry degree and knowledge of biology and physics throughout his previous careers, he was more qualified than any of the other experts. But his conclusions still should have been scrutinized.

"One can tell, for example, if the blood spattered quickly or slowly, but some experts extrapolate far beyond what can be supported," warned the National Academy of Sciences report in 2009. "The uncertainties associated with bloodstain pattern analysis are enormous."

Oscar's bloodstain pattern tests were interpretive—the results were just his opinion, not fact. In contemporary criminal cases many experiments in BPA are conducted not by forensic scientists with experience in fluid dynamics and complex math but by law enforcement officers with no scientific training or applicable credentials. The NAS report warned that only qualified experts must conduct BPA experiments. They should then accurately explain those variables when their conclusions are presented in court. Bloodstain pattern results should still not be used as the *sole evidence* in a criminal case.

"In general, the opinions of bloodstain pattern analysts are more subjective than scientific," read the report.

BPA is just too unreliable to risk a suspect's freedom . . . or life. In David Lamson's case, there was no satisfactory conclusion. Even if Oscar had conducted a true experiment, a fall against a washbasin that had actually *killed* a volunteer, there would have still been too many variables. The human body can be unpredictable.

And there is another valid theory—one that might have explained both Oscar's conclusion that Allene died by striking the sink and the prosecutor's argument that her death was murder. If David Lamson had killed his wife during a fit of rage, a spontaneous attack, why would he have brought a pipe from the backyard with him? What if he had actually thrown her onto the sink basin himself, producing the deadly fracture and the curved, parallel lacerations? It was a theory that no one presented in court—but the evidence proved it was possible.

Allene Lamson's death would forever remain an enigma. It was impossible to say for certain whether Oscar Heinrich had helped free an innocent man . . . or helped free a murderer.

EPILOGUE

Case Closed

By 1953, Oscar Heinrich's business, along with those of many experts, had survived the nation's most turbulent era, and forensic science flourished. The following decades saw the development of computerized fingerprint scans, advances in toxicology, and DNA testing—all sciences designed to help law enforcement solve cases.

Some of Oscar Heinrich's contemporaries went on to become historical figures in the history of criminal law. August Vollmer, Oscar's colleague at UC Berkeley, became a leader in the development of the field of criminal justice. Calvin Goddard was responsible for a number of important advancements in ballistics. Dr. Paul Kirk helped Vollmer establish University of California at Berkeley's School of Criminology, a program available to anyone who wanted a criminology degree.

Edward Oscar Heinrich became one of the most influential forensic scientists in history. Historians and young investigators have studied his celebrated career, and many of his investigations literally became textbook cases that students would examine for decades to come.

A generation after the criminalist used handwriting to profile a priest's murderer, the federal government trained their agents on how to identify a killer's habits to reveal his identity.

In the years after Oscar pioneered forensic geology in the case of

Father Heslin, it became common practice in solving crimes and convicting murderers.

His method of photographing evidence in the Martin Colwell case using a comparative microscope was revolutionary, a shift forward for forensic ballistics. That method is also still utilized today.

Oscar's use of forensic entomology in the Bessie Ferguson case—the first in America—is still a powerful weapon for modern investigators. Insects have starred in countless investigations since Oscar's landmark case in 1925.

His deductive reasoning in the Siskiyou train robbery is still called one of the most astounding examples of how to analyze trace evidence. Forensic historians credit Oscar with developing the first method of meticulously sorting, categorizing, and cataloguing evidence—a better way to be an organized investigator. Oscar's methodology included defining what happened, where it happened, and when (in what order) it happened. His approach has since been used by thousands of investigators.

Oscar's stories from the field and inside the lab astounded his college students each semester. And that is perhaps his greatest legacy—Oscar taught criminology classes at UC Berkeley for almost thirty years. Thousands of students learned from him, studied his techniques, and then used those tools in their own careers. As a sought-after professor in the most popular criminology school in the country, Oscar was one of criminal justice's greatest influencers.

While Oscar Heinrich innovated some of the most groundbreaking methods in forensics, he also advanced some dubious methods, including handwriting analysis and bloodstain pattern analysis—both now considered junk science. In fact, they can be quite dangerous in the courtroom when used as the main evidence, and both methods are becoming less frequently used by investigators to support their theories in criminal cases. Unfortunately, Oscar would not be the last innovator of unreliable science.

In 1932, the FBI's first director, J. Edgar Hoover, created the federal crime laboratory and began hiring many of its own experts and special agents, which reduced the need for independent forensic scientists like Oscar Heinrich. But in 1997, the Justice Department criticized the FBI lab for using flawed scientific practices, like comparative bullet-lead analysis, that might have tainted dozens of criminal cases. "The investigation found that the laboratory's explosives, chemistry-toxicology and materials analysis units were rife with substandard performance that had forced F.B.I. officials to review several hundred past and current cases to determine how many might have been jeopardized by faulty work," reported the *New York Times*.

Oscar Heinrich warned that sloppy work from poorly trained scientific examiners would undermine criminal justice, and he was right. Recent studies conducted by the Innocence Project have shown that misguided lab technicians or bad science techniques, like bite mark evidence or shoe print comparisons, were responsible for almost 50 percent of wrongful convictions.

The 2009 report from the National Academy of Sciences offered scathing evidence that crime labs across America needed overhauling, including standardized techniques and better-trained examiners. It targeted much-used forensic disciplines like fingerprinting, firearms identification, bite marks, and BPA, along with handwriting and hair analysis—all tools used for decades to catch criminals.

The training of experts can vary wildly from agency to agency, creating a lack of consistency that should be frightening to anyone involved in a criminal case, according to the NAS. There is also no consistency in their certification or in the accreditation of crime laboratories. The majority of jurisdictions don't require forensic practitioners to be certified, or even formally educated in these techniques. And most forensic science disciplines have no mandatory certification programs. Judges on cases involving bloodstain pattern analysis have allowed "experts" to take the stand whose credentials have included only a forty-hour course.

Another issue, according to the report, is that state and local law enforcement agencies sorely lack resources like funding, staff, and equipment "to promote and maintain strong forensic science laboratory systems."

The NAS recommended many changes, including rigorous independent testing of all forensic techniques that currently have little scientific support, which are virtually all of them except DNA and toxicology. Much of forensic science was developed within law enforcement agencies, and the techniques are not subjected to the rigorous, systematic tests and peer reviews like other scientific techniques. That *has* to change to maintain the integrity of the criminal justice system.

The academy suggested that the advent of an independent federal agency would standardize forensic science by financing research and training, hoping to join more closely the pursuit of truth by both the scientific community and law enforcement. An aggressive initiative would take tremendous support from Congress and the White House—as of 2020, that is unlikely to happen.

Attorneys are more often requesting Daubert Hearings to test the credentials of experts. Judges are asked to evaluate the admissibility of an expert, a forensic method, testimony, or evidence while the jury sits nearby. But judges are generally not versed in scientific techniques, and they are ultimately the gatekeeper. Junk science is still admitted to trial.

In 1992, Cameron Todd Willingham was convicted in Texas for setting a fire that had killed his three young daughters—a two-year-old and one-year-old twins. Willingham had always maintained his innocence, but police investigators claimed that they could prove that the fire had been started by some form of liquid accelerant. Almost twenty years later, leading arson experts concluded that the evidence was unreliable, and further evidence in the case exonerated Willingham. But it was too late—Cameron Todd Willingham had been executed in 2004.

Shoddy forensic techniques can also lead to far-reaching tragedies. In 2014, a former crime scene analyst committed suicide in California after his DNA was erroneously connected to the sexual assault and murder of a fourteen-year-old girl in 1984. After Kevin Brown's death, a federal judge ruled that his DNA was likely present on the victim's body because of cross contamination, probably due to now-outdated standards used in his crime laboratory. Brown died believing that he would forever remain a murder suspect.

We can still learn so many crucial lessons from the nascent development of forensic science. Investigations must start with honest, intelligent officers willing to do good detective work in the field. The public should question law enforcement without impeding its progress, and jurors shouldn't be swayed by an expert's reputation—they should evaluate if his theory makes sense. America should continue to lead the way in the development of forensic science through federal funding and research. Confessions should not be enough to convict. In fact, false confessions are at least partially responsible for more than 25 percent of wrongful convictions, while eyewitness misidentification is another leading cause.

All forensic science is fallible, even DNA testing. Americans can only hope that investigators will doggedly gather reliable evidence, clues that can get to the truth rather than settle on an outcome that will appease the public or free a guilty suspect. Investigators in the field and inside the laboratory must continue to develop their arsenal through meticulous research in forensic science, psychology, and profiling, and then those methods need to be rigorously tested against scientific standards for reliability. If they fail, then little stands between criminals and the rest of society. And even one innocent person behind bars means the judicial system has failed.

Oscar Heinrich never forgot the David Lamson case, one of the most controversial of his career. In the end, the scientific evidence confused jurors more than anything. David Lamson's life was spared not because he was innocent but because some of the nation's top forensic experts could not agree on the evidence.

Oscar would never be fully victorious over the skepticism of forensics in a courtroom, but he had helped jurors get much closer to understanding how scientists in lab coats could protect society.

"Don't know whether I've been tossed into the dangerous seventies as a wolf, a cripple or a eunuch," Oscar Heinrich told John Boynton Kaiser in 1950, several months before his seventieth birthday. "The end result is promised somewhere between the first and the second condition. Can't complain, having nearly made my allotted three score and ten."

Oscar Heinrich had spent the two decades since David Lamson's trials solving as many as one thousand more cases, ranging from murders, to forgeries, to fabricated wills. He watched his two children grow into strong men who still strived to impress their demanding father. Mortimer and Theodore both survived bombings and sniper fire during World War II. Mort received a Purple Heart when he took a hit during the Battle of Leyte in the Pacific campaign, and then he was awarded a Bronze Star. Theodore also earned a Bronze Star, along with other commendations. But perhaps Theo's most impressive achievement during the war was his assignment as one of the famous Monuments Men, a group largely of art historians and museum personnel from fourteen Allied nations who identified and returned art stolen by the Nazis.

Both of his sons followed Oscar's own career path, albeit briefly. Mortimer studied criminalistics and became a handwriting expert; father and son worked together briefly on questioned-documents investigations until Mort settled into a career at the Bank of Hawaii in Honolulu. In the early 1940s, Theo and Oscar traveled through Europe together as art authenticators, but eventually Theo became Associate Curator of Paintings at the Metropolitan Museum of Art in New York and later became a professor of art history at York University in Toronto.

The strain between Theodore and Oscar over their mutual money problems never quite resolved itself, because even though Theo was working for the Metropolitan Museum of Art, the forty-three-year-old still requested money from his father. The elderly criminalist told Theo just a month before his death that Oscar could not retire until the leaks were plugged.

"I am still waiting for you to put your financial commitments in good order, and of your own volition," Oscar told Theo in August 1953. "There is a time when you were broke and needed much sympathy and encouragement, but those days happily should be over."

Oscar had suffered from a serious stroke the previous October, surely a sign that scaling back would have been wise. But the seventy-two-year-old, weak from waning stamina and years of chronic hypertension, ignored the warning and continued to toil in his lab for up to fifty-five hours a week. He was forced to close his San Francisco office because of high rent. He dreamed of vacations, but he felt burdened by financial obligations. Oscar was determined to stay in the laboratory to give his wife, Marion, financial security.

"I am annoyed by nothing other than your Mother's anxiety over the attention that I continue to give my working program," he told Theo. "Recently I have put handrails on the steps from the lower levels to the living room level."

As his sons continued to advance in their own careers, Oscar offered them advice he had gathered from onerous lessons, particularly those delivered by his rival experts. Time had not mended those relationships, and competition had stoked their mutual contempt.

"You will find as you grow in your field that as you meet with success there will always be some who will believe that you reached your upper level through graft and special favors rather than ability," Oscar told Mortimer. "Pay no attention to it, just keep on growing."

He enjoyed being a grandfather to Mortimer's three children, and while he lamented never being a successful novelist, he did seem content with publishing useful articles and books about forensic science. Oscar absolutely adored Theo and Mortimer despite decades of complaints about their lack of financial responsibility. He was disappointed that neither son took an interest in taking over the family business, but he also settled on the idea of leaving behind a different legacy.

"I regard my greatest accomplishment to be the completed preparation of two sons, each able and willing to contribute to the welfare of the community in which they live and work," Oscar wrote Theo. "It has been a long pull for Dad, but one in which I never for a moment have entertained a doubt as to the outcome."

Working alone in his laboratory late at night on September 23, 1953, Edward Oscar Heinrich suffered a second stroke. He was treated by doctors for five days as his family surrounded him in the hospital. His best friend, John Boynton Kaiser, dashed off a sentimental telegram.

"Best wishes. Get well soon," Kaiser said on September 29. "In spite of your fine achievements the world still needs you and your friends hope you will soon be adding new laurels to the record."

But the note arrived too late—Oscar had died the day before without ever regaining consciousness. And in some ways, the legend of "America's Sherlock Holmes" died with him. Oscar Heinrich's fingerprints cover the history books of forensic science because so many of his techniques are still used. Experts today call Oscar "America's greatest forensic scientist of the early twentieth century."

"Almost single-handedly, he restored the reputation of the expert witness in the American courtroom," wrote one forensics writer. "With uncanny feats of deduction that seemed to spring from the pages of fiction, Heinrich acquired legendary status, earning a forensic celebrity that extended far beyond his homeland."

Oscar never pursued the spotlight—it chased him. A shining light in the dark world of crime, his amazing feats in criminal investigations were unmatched during his time, and that's likely true today. Rarely now is there an investigator with his expertise in so many disciplines—someone who can synthesize those skills with outstanding fieldwork and deductive reasoning in the laboratory. And that's tragic, because there might be some challenging cases, a few impossible modern mysteries, that would require an American Sherlock Holmes like Edward Oscar Heinrich to solve them.

ACKNOWLEDGMENTS

One of the most exciting things about this book, as the author, is how it was created. It nearly failed before it began. Several years ago, I read about the Siskiyou train robbery in an encyclopedia of American crime (I bet you never knew one of those existed!) and noticed Edward Oscar Heinrich's moniker, "America's Sherlock Holmes." How could a true crime author ignore *that* description?

I was on the hunt for my second book, and writing a biography seemed intriguing. I discovered Heinrich's collection at UC Berkeley and was thrilled to find out that it is very, very large. Too large. In fact, the university had avoided archiving it because it was so vast, so unwieldy. There was a form on the website to request a review from an archivist, sort of like a plea from the public. It required that I explain *why* I needed this particular collection organized and then made available. I explained that I'm an associate professor of journalism at the University of Texas at Austin and a nonfiction author. And I argued that Edward Oscar Heinrich was one of the most prolific forensic scientists in American history. He deserved to have a book written about him.

I sent the form . . . and then waited for two months. Finally, an assistant at the Bancroft Library emailed me with horrible news—they were *not* planning to archive Heinrich's collection anytime soon. They were understaffed and there were other priorities. And then the assistant said something akin to: "You do realize just how *big* this collection is?" I did. But I just knew that this was *my* book. I emailed the archivist again, a few weeks later. "Will they reconsider?"

Lara Michels, the library's head of archival processing, responded a day or two later with surprising news. "Yes. We'll do it." She had looked through some of the materials and agreed that Edward Oscar Heinrich was extraordinary. She would process the collection herself.

The boxes in Heinrich's collection, more than one hundred, were all stored in an off-site facility not far from the library. Michels would dedicate one day a week for the next eighteen months to traveling there and cataloguing his collection. I was elated but also a bit worried that the process would be painfully slow. And it certainly was. But Michels's work on Heinrich's collection was phenomenal. She catalogued every tiny, itty-bitty thing she discovered. She meticulously read through each document; she squinted at each piece of evidence that Heinrich kept (which should have been turned over to the police by him, frankly). She dismissed nothing.

Michels would email me photographs of interesting discoveries, things that only I would appreciate, just so I could understand this man a little bit better. When she was finally through cataloguing the collection, she printed out a special guide, just for me. And she invited me to sit with her inside the off-site facility so I could get an early start on my research, something she's never done before. She continued to organize Heinrich's boxes as I photographed his lifelong work in criminal investigations. We talked about his family, his life, and his cases. She had advised me on which investigations to include in the book because those files were the most robust. I told her about finding trial transcripts for each of his cases. I showed her letters from Theodore Heinrich's own collection at the University of Regina in Canada. She suggested that I search two other collections at UC Berkeley: those of John Boynton Kaiser and August Vollmer. Michels marveled at the photos of the David Lamson case that I found at Stanford University. I emailed her when I discovered August Heinrich's suicide.

There's no better relationship in the world of publishing, I think, than the one between an author and an archivist who both *truly* want a book to succeed. Edward Oscar Heinrich and his work would be lost in history were it not for Lara Michels and UC Berkeley. I am eternally grateful to both.

There are some other folks who deserve accolades:

My incredible fact-checker, Joyce Pendola, was an able safety net, as usual. Former defense attorney and law professor David Sheppard is always a fantastic listener and adviser. Outstanding reporter and good friend Pamela Colloff vetted my sections on bloodstain pattern analysis, an incredibly confusing discipline. University of Texas psychology professor Kim Fromme helped me pin down Oscar Heinrich's odd characteristics. Daniel Wescott, who is the director of the Forensic Anthropology Center at Texas State University, advised me on the different ways that bodies can decompose (his guidance was needed in several chapters).

Drs. Jill Heytens and Steven Kornguth, both well-regarded neurologists, advised me on the Allene Lamson case. Dr. Heyens even volunteered to help me reenact Allene's fatal accident. We decided against it. Huge thanks to Tina Shorey and Desi Rodriguez, who escorted me to a shooting range and then wisely stepped out of the way as I learned how to fire several large guns.

I would be remiss not to thank some folks at the University of Texas, particularly the dean of the Moody College of Communication, Jay Bernhardt, and School of Journalism director Kathleen McElroy. I've received generous funding for several years from the Maureen Healy Decherd '73 Teaching Endowment for Journalism, for which I am grateful. I'm also so happy to see my cousin Diana Dawson right across the hall from me at UT. She inspired me to become a journalist years ago. And I'm so pleased to share students with the most talented journalism faculty in the country—their endless support is truly wonderful.

Thanks to Becka Oliver, the executive director of the Writers' League of Texas, who was a fantastic cheerleader from the beginning—and her organization is outstanding. Emily Donohue at Stratfor has been so supportive.

I'm honored to be with G.P. Putnam's Sons for the first time. I'm absolutely astounded by the amount of talent on this team. I'm blessed to have been surrounded with a passionate group of people who quite honestly love and support this book, particularly Ivan Held, president; Sally Kim, editor in chief; Katie Grinch, associate director of publicity; Ashley McClay, director of marketing; Brennin Cummings, assistant marketing manager; and Gabriella Mongelli, assistant editor. I will forever be indebted to executive editor and good friend Michelle Howry—a gifted editor who can help craft a narrative into something truly wonderful. I'd be lost without her.

This book would still be a sad, forgotten file on my hard drive were it not for my literary agent, Jessica Papin, with Dystel, Goderich & Bourret. There's no one in this wild world of books whom I trust more.

To my Texas girls, my closest friends of thirty years—I continue to write books only as an excuse to throw a boat party with you on Lake Travis.

To my late father, Robert Oscar Dawson, a brilliant professor of law at the University of Texas for thirty-seven years. He taught me from an early age that one wrongly convicted person is too many.

And finally, to Jenny, Ella, and Quinn; as well as my parents, Lynn and Jack Lefevre; my in-laws, Sandra and Charlie Winkler; and my brothers-in-law, Chuck Winkler and Shelton Green—you all keep my ship sailing along.

NOTES

PROLOGUE

1 **His upper jawbone was massive:** Bessie Ferguson case, 1925, carton 24, folder 5, Edward Oscar Heinrich Papers, BANC MSS 68/34 c, Bancroft Library, University of California, Berkeley.

2 **People with OCPD:** Encyclopedia of Mental Disorders, http://www.minddisorders.com; the expertise of Dr. Kimberly Fromme, professor of Clinical Psychology at the University of Texas at Austin.

3 **Every conceivable type of microscope:** From a photo in 89–44, box 175, file 2291, Theodore Heinrich Collection, Dr. John Archer Library, University of Regina, Saskatchewan, Canada.

4 ***Blood, Urine, Feces and Moisture: A Book of Tests*:** This title and other examples from Heinrich's library come from the University of California at Berkeley library catalogue: search "Edward Oscar Heinrich."

4 **increased by as much as almost 80 percent from the decade before:** Mark Thornton, "Policy Analysis No. 157: Alcohol Prohibition Was a Failure," Cato Institute, July 17, 1991 (specifically the section "Prohibition was criminal"). Note: the numbers for the increase in crime are widely disputed, ranging anywhere from 5 percent to 78 percent.

4 **Summary of crime in the 1920s:** Encyclopedia.com, "Crime 1920–1940."

4 **The FBI was still the Bureau of Investigation:** History of the FBI, Federal Bureau of Investigations, https://fas.org/irp/agency/doj/fbi/fbi_hist.htm.

4 **especially women, whose newfound independence:** "'Sex Appeal' Responsible for U.S. Crime Wave," *Times of India,* March 22, 1926.

5 **"Footprints are the best clue":** "Foot-Print Is the Best Clue," *Casper Star Tribune* (WY), January 17, 1928.

5 **one of America's greatest forensic scientists:** Katherine Ramsland, "He Made Mute Evidence Speak: Edward O. Heinrich," *Forensic Examiner* 16, no. 3 (Fall 2007): 2.

5 **In 1910, when he opened the nation's first private crime lab:** Max M. Houck, ed., *Professional Issues in Forensic Science* (Amsterdam: Elsevier, 2015), 3; carton 5, folder 5, Edward Oscar Heinrich Papers.

5 **"America's Sherlock Holmes":** "Grains of Sand Convict Killer," *Reno Gazette-Journal,* February 18, 1930.

5 **nascent innovator of criminal profiling:** Tom Bevel and Ross M. Gardner, *Bloodstain Pattern Analysis,* 3rd edition (Boca Raton, FL: Taylor & Francis, 2008).

Notes

CHAPTER 1

7 **Her husband was fond of burning the rubbish:** Criminal No. 3730, *In the Supreme Court of the State of California, the People of the State of California vs. David Lamson* (San Francisco: Pernau-Walsh Print. Co., October 1934), 114.

7 **David Lamson in the yard:** Ibid., 112–18.

7 **garden trimmings, dead artichoke plants:** Ibid., 117, 247.

8 **palatial homes of professors and professionals:** Theresa Johnston, "These Old Houses," *Stanford Magazine,* November/December 2005.

8 **the fourth year of the Great Depression:** "Great Depression History," History .com, October 29, 2009, https://www.history.com/topics/great-depression/great-depression-history.

8 **Herbert Hoover's impressive three-tiered residence:** "Was 'Bathtub Murder' an Accident?," *Decatur Herald* (IL), July 1, 1934.

9 **Professors and scholars at Stanford University continued to teach classes:** "History of Palo Alto," https://www.cityofpaloalto.org/gov/depts/pln/historic_preservation/history_of_palo_alto.asp.

9 **Allene was a natural beauty:** Per photos contained in the Lamson Murder Case Collection (SC0861), 1933–1992, Department of Special Collections and University Archives, Stanford University.

10 **"In a few short miles one passes from sea level":** "The Campus as a Game Refuge" in "Clippings, articles, publications" in ibid.

10 **engaged in the Stanford community:** Ibid.

10 **David was the sales manager:** *Supreme Court of the State of California,* 14–15.

10 **He had spent a year teaching advertising:** "The 1930s: Continued Growth," *Stanford: 125 Years of Journalism,* http://www.125yearsofjournalism.org/1930s.

10 **"She needed something to occupy her mind":** *Supreme Court of the State of California,* 258.

10 **The Lamsons were a modish couple:** *Supreme Court of California statement PEOPLE v. LAMSON,* 1 Cal.2d 648 (San Francisco: Pernau-Walsh Print. Co., October 1934), 1.

11 **moneyed figures in Palo Alto:** Ibid., 17.

11 **"I would say they were quite happy":** *Supreme Court of the State of California,* 111.

11 **Details from this weekend:** Various sources, including ibid., 28–29, 364–65.

12 **horrible sinus infections:** Ibid., 257–58.

12 **Much of the tiny bathroom was bright white:** "Experiment Summary," dated June 20, 1933, in the "Lamson" folder found in carton 71, folder 31–41, Edward Oscar Heinrich Papers.

13 **she suffered from notoriously weak ankles:** Frances Theresa Russell and Yvor Winters, *The Case of David Lamson* (San Francisco: Lamson Defense Committee, 1934), 30.

13 **Allene Lamson's "physical exam records" as a student:** Lamson Murder Case Collection.

13 **The tub was about halfway:** "Testimony of Chief H. A. Zink Given at Second Trial," Lamson Murder Case Collection, 7.

13 **The doorbell rang:** *Supreme Court of the State of California,* 121.

13 **the sensation was startling:** Details about Allene Lamson's likely physical reaction to her injuries come from Dr. Jill Heytens, a neurologist with almost thirty years in practice.

13 **Her torso dangled halfway out:** Description of Allene's body, including her braids, comes from a crime scene photo in Lamson Murder Case Collection.

14 **"I hoed":** *Supreme Court of the State of California,* 117.

14 **"I remarked that he was doing":** Ibid., 112.

15 **"He said it would be perfectly":** Ibid., 121.

15 **"I really cannot describe it":** Ibid., 130.

15 **"My God, my wife has been murdered!":** Russell and Winters, *The Case of David Lamson*, 49. Note: there was much debate over that statement. Mrs. Place, the real estate agent, testified that he said "murdered" rather than "killed." Her client, Mrs. Raas, testified he said "killed."

15 **"The first thing I saw was blood":** David Lamson, *We Who Are About to Die* (New York: Charles Scribner's Sons, 1936), viii.

15 **"Of the rest of that morning I remember":** Ibid.

16 **"Get the police to find the murderer!":** *Supreme Court of California statement PEOPLE v. LAMSON*, 2.

16 **One neighbor said:** "Husband Is Held When Wife Killed," *Healdsburg Tribune* (CA), May 30, 1933.

16 **"Some of the things I remember most vividly":** Lamson, *We Who Are About to Die*, viii–ix.

16 **his neighbor Mrs. Brown found him:** "Lamson Sits Silent as He Hears Charges," *Santa Cruz News* (CA), June 16, 1933.

16 **"The glimpse of a neighbor's face":** Lamson, *We Who Are About to Die*, xi.

16 **"She was down, cleaning up something off the floor":** "Testimony of Chief H. A. Zink," 2–3.

16 **Eight officers had responded:** *Supreme Court of the State of California*, 138.

16 **and the autopsy later concluded that Allene had died:** Ibid., 29.

17 **"Who could have done it?"** *Supreme Court of California statement PEOPLE v. LAMSON*, 2.

17 **Allene's blood had been transferred:** *Supreme Court of the State of California*, 131; "Experiment Summary," dated June 20, 1933, in the "Lamson" folder found in carton 71, folder 31–41, Edward Oscar Heinrich Papers.

17 **half of her blood had drained from her body:** Russell and Winters, *The Case of David Lamson*, 36.

17 **"Ten minutes after the deputies arrived":** Lamson, *We Who Are About to Die*, ix.

18 **"Sheriff William Emig expressed":** "Prominent Young Palo Alto Woman Is Found Dead in Bath Tub with Gaping Hole in Back of Her Head," *Santa Cruz News* (CA), May 30, 1933.

18 **kidnapped for ransom from the family's mansion:** "Lindbergh Kidnapping," FBI .gov, https://www.fbi.gov/history/famous-cases/lindbergh-kidnapping.

18 **"Mystery Man Adds New Theory Puzzle":** "Mystery Man Adds New Theory Puzzle," *Madera Daily Tribune* (CA), June 1, 1933.

18 **Lamson's status in the jail:** Ibid.

19 **Photos of Allene and Bebe were taped to the walls:** This and other details about his time in the San Jose jail are from Lamson, *We Who Are About to Die*, 10–11; physical descriptions of the jail originated from various photos of Lamson's cell in different collections.

19 **steadfastly believed in his innocence:** "Steadfast and True," *Oakland Tribune*, October 1, 1933.

20 **"It never occurred to any of us that anything":** Lamson, *We Who Are About to Die*, ix.

20 **a brunette was prone:** This description of Heinrich's photo shoot with Weber comes from his photos in the Lamson Murder Case Collection. The description of experiments is from "Experiment Summary," dated June 20, 1933, in the "Lamson" folder found in carton 71, folder 31–41, in Edward Oscar Heinrich Papers.

20 **"The door is liberally spattered below the glass":** Ibid.

20 **introducing perhaps the first blood-pattern analysis:** R. H. Walton, *Cold Case Homicides: Practical Investigative Techniques* (Boca Raton, FL: CRC Press, 2014), 23.

21 **"A precise little man":** "Lamson Aide Hits State; Pipe Clean," *Press Democrat* (Santa Rosa, CA), September 12, 1933.

21 **erected the nation's first private science laboratory in 1910:** Houck, *Professional Issues in Forensic Science,* 3.

22 **"Not Sherlock Holmes":** Eugene Block, *The Wizard of Berkeley* (New York: Coward-McCann, 1958), 28.

22 **"I have discovered enough evidence to:"** "Not Guilty to Be Plea by Lamson," *Santa Cruz News* (CA), June 21, 1933.

22 **"*X* marks the spot":** Nancy Barr Mavity, "Two Criminologists Reveal Evidence that Convinced Them of Lamson Innocence," *Oakland Tribune*, February 13, 1934.

CHAPTER 2

25 **Albertine had been just twenty years old:** This and other details about the Heinrich family background from http://dahlheimer-bebeau.com/Heinrich/RM-HeinrichKlemm/b98.htm#P121.

26 **"We kids earned our pennies":** This and more personal details and other quotes from Block, *The Wizard of Berkeley*, 29–34.

26 **penned a newspaper story:** Letter from Kaiser to Jacqueline Noel, May 2, 1946, box 2, John Boynton Kaiser Papers, BANC MSS 75/48 c, Bancroft Library, University of California, Berkeley.

28 **"No reason for the deed":** "Suicide at Glendale," *Tacoma Daily News*, October 7, 1897; "With the Aid of a Rope," *Seattle Post-Intelligencer*, October 8, 1897.

29 **"Among my earliest recollections":** Letter from Heinrich to Kaiser, October 9, 1922, box 12, folder 27, Edward Oscar Heinrich Papers.

29 **He studied at night to become a pharmacist:** "Memoranda of Experience," carton 4, Edward Oscar Heinrich Papers.

30 **History of pharmacy:** Joseph Fink, "Pharmacy: A Brief History of the Profession," *The Student Doctor Network*, January 11, 2012.

30 **"A drugstore is a veritable laboratory":** Block, *The Wizard of Berkeley*, 29–34.

30 **"I had doctors' prescriptions to decipher":** James Rorty, "Why the Criminal Can't Help Leaving His Card," *St. Louis Post-Dispatch*, November 9, 1924.

31 **"I was impressed with the difference":** Block, *The Wizard of Berkeley*, 32.

32 **"It may seem delightful":** Letter from Heinrich to Marion, May 12, 1932, 89–44, box 23, file 179, Theodore Heinrich Collection.

32 **Oscar and Marion were married:** "College Romance Brings Wedding," *San Francisco Call*, August 27, 1908.

32 **Heinrich's professional background:** *The American City*, volume XX (New York: The Civic Press, January–June 1919); "There Is a Destiny," *Who's Who in America*, February 1926; various files contained in series 3, Edward Oscar Heinrich Papers.

34 **Background on August Vollmer:** Frances Dinkelspiel, "Remembering August Vollmer, the Berkeley Police Chief Who Created Modern Policing," *Berkeleyside*, January 27, 2010.

34 **truth serum called scopolamine:** "Getting Confessions by New Truth Serum," *Baltimore Sun*, August 12, 1923.

35 **Teaching experience in UC Berkeley:** Document titled "Memoranda of Experience," found in carton 3, Edward Oscar Heinrich Papers.

35 **"Your main job as a cop":** Jeremy Kuzmarov, "What August Vollmer, the Father of American Law Enforcement, Has to Teach Us," *HuffPost*, October 4, 2016.

35 **America's first "cop college":** August Vollmer and Albert Schneider, "School for Police as Planned at Berkeley," *Journal of Criminal Law and Criminology* 7, no. 6 (1917): 877–98.

36 **"The course extends over a period of three years":** Harold G. Schutt, "Advanced Police Methods in Berkeley," *National Municipal Review*, volume XI (1922), 81.

36 **"I'll take you all from a thrill to a shudder":** Letter from Heinrich to Kaiser, June 30, 1920, box 1, John Boynton Kaiser Papers.

36 **"But a chess board in police station":** Letter from Heinrich to Kaiser, June 23, 1921, box 1, John Boynton Kaiser Papers.

36 **"The investigation of crime is merely a special case":** Ibid.

38 **Boulder, Colorado's first city manager:** "Boulder Appoints a City Manager," *Evening Star* (Independence, KS), February 19, 1918.

38 **Kaiser's background:** Donald G. Davis Jr. and John Mark Tucker, *American Library History* (Austin: University of Texas Press, 1978), 280.

39 **"On page 363 of the LIBRARY JOURNAL is a list of ten tests":** Letter from Kaiser to Heinrich, April 22, 1931, box 1, John Boynton Kaiser Papers.

39 **"Sometimes I enjoy your insistence in thinking":** Letter from Heinrich to Kaiser, October 31, 1921, box 1, John Boynton Kaiser Papers.

39 **the Hindu Ghadar Conspiracy:** Block, *The Wizard of Berkeley*, 49–52; Ramsland, "He Made Mute Evidence Speak."

40 **"I am enclosing a print showing the hairs":** Letter from Heinrich to Kaiser, July 9, 1921, box 1, John Boynton Kaiser Papers.

40 **"Whenever you are ready send":** Letter from Heinrich to Kaiser, July 19, 1946, box 28, John Boynton Kaiser Papers.

41 **"I now use three rooms":** Letter from Heinrich to Kaiser, May 10, 1921, box 1, John Boynton Kaiser Papers.

41 **"When I bought the car recently it":** Letter from Heinrich to Kaiser, September 15, 1921, box 1, John Boynton Kaiser Papers.

41 **"Bankers Trust Co has foreclosed mortgage":** Telegram from Kaiser to Heinrich, April 5, 1918, box 28, John Boynton Kaiser Papers.

42 **"the foremost scientific investigator of crime in the United States":** Letter from August Vollmer to Alfred Adler, July 12, 1930, box 31, August Vollmer Papers, BANC MSS C-B 403, Bancroft Library, University of California, Berkeley.

CHAPTER 3

44 **Background on Colma and its streetcars:** John Branch, "The Town of Colma, Where San Francisco's Dead Live," *New York Times*, February 5, 2016; John Metcalfe, "Remembering San Francisco's Ornate 'Funeral Streetcars,'" *CityLab* (Ridge-Field, NS), January 24, 2017; Terry Hamburg, "All Aboard! Getting to Cypress Lawn in Style Back in the Good Old Days," Cypress Lawn Heritage Foundation.

44 **Background of Father Heslin:** Jean Bartlett, "Modern-day Polygraph Dates Back to 1921 Murder in Pacifica," *Mercury News* (San Jose, CA), March 12, 2013; Heslin's physical description comes from archival photos.

44 **just ten days:** Norma Abrams, "Father Heslin's Housekeeper Accounts for Priest's Auto," *San Francisco Chronicle*, August 6, 1921.

45 **Summary of Heslin's disappearance:** People v. Hightower, Court of Appeal of California, First Appellate District, Division One, January 18, 1924, 65 Cal. App. 331 (Cal. Ct. App. 1924).

45 **Tuberculosis was the leading cause of death:** "Early Research and Treatment of Tuberculosis in the 19th Century," Historical Collections at the Claude Moore Health Sciences Library, http://exhibits.hsl.virginia.edu/alav/tuberculosis.

46 **the unemployment rate had more than doubled:** This figure references a graph in David R. Weir, "A Century of U.S. Unemployment, 1890–1990: Revised Estimates and Evidence for Stabilization," *Research in Economic History* 14 (1992): 301–46.

47 **national crime rate had increased by almost 25 percent:** Thornton, "Alcohol Prohibition Was a Failure," specifically the section "Prohibition was criminal."

47 **Crime in the 1920s:** *Encyclopedia Britannica*, "Crime 1920–1940."

47 **Bureau of Prohibition "deputized" Ku Klux Klan members:** Kat Eschner, "Why the Ku Klux Klan Flourished Under Prohibition," *Smithsonian Magazine*, December 5, 2017.

47 **arrests for drunken driving rose by 80 percent:** Thornton, "Prohibition Was a Failure," specifically the section "Prohibition was criminal."

47 **The Newton Boys:** Patricia Holm, "Newton Boys," Texas State Historical Association, June 15, 2010.

47 **Description of kidnapping and ransom notes:** John Bruce, "The Flapjack Murder," in *San Francisco Murders*, ed. Allan R. Bosworth and Joseph Henry Jackson (New York: Duell, Sloan and Pearce, 1947), 213–18.

48 **his eyes followed the curve of the letters:** Photos of ransom notes and Heinrich's analysis found in carton 70, folder 75–77, Edward Oscar Heinrich Papers.

49 **authorized an aerial search:** "Airplane to Help Authorities in Efforts to Locate Father Heslin," *Santa Cruz Evening News* (CA), August 4, 1921.

50 **"Was Priest Kidnaped to Wed Pair?":** "Was Priest Kidnaped to Wed Pair?," *Oakland Tribune*, August 4, 1921.

50 **"I have nothing further to say":** Abrams, "Father Heslin's Housekeeper Accounts for Priest's Auto."

50 **History of handwriting analysis:** Forensic Document Examination: A Brief History, National Institute of Standards and Technology.

51 **forged will case in Montana:** "Murray Will Is Forgery Is Opinion of Handwriting Experts of California," *Great Falls Tribune* (MT), June 22, 1921.

52 **A 2009 landmark study by the National Academy of Sciences:** Committee on Identifying the Needs of the Forensic Sciences Community, National Research Council, *Strengthening Forensic Science in the United States: A Path Forward* (Washington, D.C.: National Academies Press, 2009), 184.

52 **"Some cases of forgery are characterized":** Ibid., 166–67.

52 **less reputable discipline: graphology:** Russell W. Driver, M. Ronald Buckley, and Dwight D. Frink, "Should We Write Off Graphology?," *International Journal of Selection and Assessment* 4, no. 2 (April 1996): 78–86.

53 **"Notice how mine goes over the top":** Letter from Heinrich to Kaiser, January 31, 1921, box 1, John Boynton Kaiser Papers.

54 **"The writer is demented":** "Experts Believe Writer of Ransom Letter Demented," *San Francisco Chronicle*, August 6, 1921.

54 **"I am expected to do all of the heavy work":** Letter from Heinrich to Kaiser, September 6, 1921, box 1, John Boynton Kaiser Papers.

54 **But Oscar labeled the other expert:** Various letters found in carton 70, folder 75–77, Edward Oscar Heinrich Papers.

55 **representing attorneys for the American government had arrested McGovern:** United States v. Chauncey McGovern, G.R. No. 2731 (November 6, 1906).

55 **"The writer of that letter is a baker":** Block, *The Wizard of Berkeley*, 81.

56 **History of criminal profiling:** Katherine Ramsland, "Criminal Profiling: How It All Began," *Psychology Today*, March 23, 2014; "Offender Profiling," *World Heritage Encyclopedia*, 2015.

57 **"Fate has made me do this":** "$15,000 Ransom for Priest Is Asked in New Kidnap Note," *Santa Ana Register* (CA), August 10, 1921.

57 **"The handwriting shows the writer to be a jellyfish":** Bruce, "The Flapjack Murder," 222.

57 **the unfounded rumor of a third letter:** "An Old-Time Scoop in San Francisco," *Editor & Publisher*, September 10, 1921.

58 **Hightower was the stranger's name:** Bruce, "The Flapjack Murder," 222–24.

58 **Description of the archbishop's home:** "1000 Fulton Street," *Dona Crowder*, http://www.donacrowder.com/1000-Fulton-Street.

58 **"I don't know this man":** "An Old-Time Scoop in San Francisco."

58 **Palm Beach suit:** "Art Collection" section of the Metropolitan Museum of Art website.

60 **Fundamentalism revival:** "Religious Fundamentalism, Twentieth Century History," BBC article; Grant Wacker, "The Rise of Fundamentalism," National Humanities Center.

60 **bootlegging, Walgreens, and medicinal whiskey:** Evan Andrews, "10 Things You Should Know about Prohibition," History.com, January 16, 2015.

61 **The sign showed a drawing:** Bruce, "The Flapjack Murder," 225, along with a photo of Albers Mill flapjack flour found online.

62 **"If the body is in here":** Ibid., 226.

CHAPTER 4

65 **"Get Ready":** "An Old-Time Scoop in San Francisco."

65 **More details about case:** "Missing Priest Was Murdered," *Chanute Daily Tribune* (KS), August 11, 1921.

66 **"You've got a funny way of showing your gratitude":** Block, *The Wizard of Berkeley*, 86.

67 **"Father Heslin has made the supreme sacrifice":** "Father Heslin Poured Forth Own Blood in Adoration of God, Says Archbishop in Tribute," *San Francisco Chronicle*, August 14, 1921.

68 **"There is a small patch at the foot of the large blade":** This and reports from Heinrich's tests come from carton 70, folder 75–77, Edward Oscar Heinrich Papers.

68 **"Now I'm going to work on the other things":** Block, *The Wizard of Berkeley*, 86–93; and carton 70, folder 75–77, Edward Oscar Heinrich Papers.

69 **petrographic analysis:** James Gregory McHone, "Polarizing, Petrographic, Geological Microscopes," May 11, 2013, http://earth2geologists.net/Microscopes/.

70 **History of forensic geology:** Alastair Ruffell and Jennifer McKinley, "Forensic Geoscience: Applications of Geology, Geomorphology and Geophysics to Criminal Investigations," *Earth-Science Reviews* 69, no. 3–4 (March 2005): 235–47.

70 **Locard's Exchange Principle:** Claude Roux et al., "The End of the (Forensic Science) World as We Know It?: The Example of Trace Evidence," *Philosophical Transactions of the Royal Society B* 370, no. 1674 (August 2015).

71 **"Five years on the brink of bankruptcy":** Carton 70, folder 75–77, Edward Oscar Heinrich Papers.

71 **1920s mental health treatment:** Phil Hickey, "Legacy of Abuse," *Behaviorism and Mental Health*, October 2, 2011; Zeb Larson, "America's Long-Suffering Mental Health System," *Origins: Current Events in Historical Perspective* 11, no. 7 (April 2018).

71 **Other mental health milestones:** Jess P. Shatkin, "The History of Mental Health Treatment," New York University School of Medicine, 21, 24, 30, 33.

71 **"I find that I have to make an engagement in court":** Letter from Heinrich to Kaiser, September 15, 1921, box 1, John Boynton Kaiser Papers.

72 **"He has hustled around to the newspapers":** Letter from Heinrich to Kaiser, September 6, 1921, box 1, John Boynton Kaiser Papers.

72 **"Were it not for the fact that this is a criminal matter":** Ibid.

73 **"Whatever she says is all right":** Bruce, "The Flapjack Murder," 233–34.

73 **John Larson background:** Ezra Carlsen, "Truth in the Machine: Three Berkeley Men Converged to Create the Lie Detector," *California Magazine*, Spring 2010.

74 **How Larson's polygraph worked:** "John Larson's Breadboard Polygraph," The Polygraph Museum, http://www.lie2me.net/thepolygraphmuseum/id16.html.

74 **Different sections for polygraph testing:** "Psychologists Called Upon to Solve Murder," *San Francisco Chronicle*, August 18, 1921.

74 **character of Wonder Woman:** Sarah Sloat, "The Bunk Science that Inspired 'Wonder Woman,'" *Inverse*, June 6, 2017.

74 **Larson's machine:** Carlsen, "Truth in the Machine."

75 **"He will be spirited out of the city":** "Story of Hightower Is Gradually Being Broken by Police," *Madera Mercury* (CA), August 14, 1921.

75 **"My God! It's him!":** "Housekeeper for Heslin Positive in Her Identification," *San Francisco Chronicle*, August 17, 1921.

75 **Eyewitness misidentification is the leading contributing factor:** "In Focus: Eyewitness Misidentification," The Innocence Project, October 21, 2008, https://www.innocenceproject.org/in-focus-eyewitness-misidentification.

75 **Use of polygraphs:** "The Truth About Lie Detectors (aka Polygraph Tests)," *American Psychological Association*, August 5, 2004.

75 **Portable test:** Kerry Segrave, *Lie Detectors: A Social History* (Jefferson, NC: McFarland & Company, 2003), 17–18.

76 **"My head seems to swell when I think":** "Hightower Preparing for Plea of Insanity," *Freeport Journal-Standard* (IL), August 16, 1921.

76 **"I'm through":** "'I Never Saw Her Before,' Hightower Declares When Faced by 'Dolly Mason,'" *Oakland Tribune*, August 16, 1921.

76 **Hightower is questioned:** "Prisoner's Nerves Shattered by Evidence Connecting Him with Priest's Murder," O*akland Tribune*, August 17, 1921.

77 **"The suspect was covering up important facts":** "Psychologists Called Upon to Solve Murder," *San Francisco Chronicle*, August 18, 1921.

77 **"Mere embarrassment or fear are registered":** "Detect Falsehoods by Blood Pressure," *Bend Bulletin* (OR), March 10, 1922.

77 **"'general acceptance' in the relevant scientific community":** National Research Council, *Strengthening Forensic Science*, 88.

77 **"underlying reasoning or methodology is scientifically valid":** D. Daubert, et al. v. Merrell Dow Pharmaceuticals, Inc., United States Court of Appeals, 9th Circuit (June 28, 1993), 593.

78 **"I have been thinking too much lately":** Bruce, "The Flapjack Murder," 237–38.

78 **George Lynn's testimony:** "Priest's Grave Sought at Eerie Midnight Hour," *Morning Register* (Eugene, OR), October 6, 1921.

78 **Prosecutor's witnesses:** "Evidence Web Tightens," *Los Angeles Times*, October 7, 1921.

78 **"There was good light":** "Hightower Abductor Is Word," *Santa Ana Register* (CA), October 10, 1921.

79 **"it would be unwise to ask that":** "Doris Shirley Putnam Shatters Alibi Story of Wm. Hightower; New Testimony Is Produced," *San Francisco Chronicle*, October 7, 1921.

80 **"Your memory is woefully short, little girl":** Bruce, "The Flapjack Murder," 237.

81 **"This evidence was considered the strongest":** "Bad Day for Hightower," *Riverside Daily Press* (CA), October 8, 1921.

81 **"Look at that letter D":** Block, *The Wizard of Berkeley*, 92–93.

82 **"Farewell to criminals and their detection":** Letter from Kaiser to Heinrich, October 22, 1921, box 1, John Boynton Kaiser Papers.

82 **"Our copy of Lucas 'Forensic Chemistry' came yesterday":** Letter from Kaiser to Heinrich, October 25, 1921, box 1, John Boynton Kaiser Papers.

83 **"It occurs to me that you may wish to write the *Literary Digest*":** Letter from Kaiser to Heinrich, September 30, 1921, box 1, John Boynton Kaiser Papers.

83 **"Your idea on articles which you have suggested is not bad":** Letter from Heinrich to Kaiser, October 31, 1921, box 1, John Boynton Kaiser Papers.

83 **"Your unflattering comments on the titles I suggested":** Letter from Kaiser to Heinrich, November 10, 1921, box 1, John Boynton Kaiser Papers.

84 **Details about verdict:** "Jury Finds Hightower Guilty of Killing Priest," *Des Moines Register*, October 14, 1921.

84 **"We find the defendant guilty of first degree murder":** "Hightower Gets Life Tomorrow," *Santa Cruz Evening News* (CA), October 14, 1921.

84 **"Well, boys, I guess":** Ibid.

85 **"In my opinion the case got away from Hightower":** Letter from Heinrich to Kaiser, December 3, 1921, box 1, John Boynton Kaiser Papers.

85 **"The knife":** Letter from Heinrich to Kaiser, October 31, 1921, box 1, John Boynton Kaiser Papers.

85 **"Regardless of what the jury and the public":** "Pastor Convicted of Murder, Says Innocent," *Lansing State Journal*, October 14, 1921.

86 **"I have no feelings, no bitterness against anybody":** "Half His Life Left in Prison, 86-Year-Old Man Goes Free," *Amarillo Globe-Times* (TX), May 24, 1965; "Priest Slayer Receives Parole After 43 Years," *Fresno Bee* (CA), March 30, 1965.

CHAPTER 5

87 **"Virginia Rappe was dying":** Greg Merritt, *Room 1219: The Life of Fatty Arbuckle, the Mysterious Death of Virginia Rappe, and the Scandal That Changed Hollywood* (Chicago: Chicago Review Press, 2013), 42.

87 **The twenty-six-year-old:** Reports of Rappe's age varied, but *Find a Grave* reports she was born in 1895, while Room 1219 says she was born in 1891.

87 **A showgirl named Maude Delmont:** "Film Tragedy Uncovers Rum 'Road' on Coast," *Washington Times,* September 19, 1921.

88 **no telltale evidence of sexual assault:** Merritt, *Room 1219,* 42.

88 **bottles of scotch, gin, wine, and bourbon:** Andy Edmonds, *Frame-Up!: The Untold Story of Roscoe "Fatty" Arbuckle* (New York: William Morrow & Co., 1991), 154; Merritt, *Room 1219,* 8.

88 **Rappe background:** Jude Sheerin, "'Fatty' Arbuckle and Hollywood's First Scandal," *BBC News*, September 4, 2011; Merritt, *Room 1219,* 42.

88 **Fatty Arbuckle's reputation:** Gilbert King, "The Skinny on the Fatty Arbuckle Trial," *Smithsonian Magazine*, November 8, 2011; Sheerin, "'Fatty' Arbuckle and Hollywood's First Scandal."

89 **History of St. Francis:** "History," Westin St. Francis, https://www.westinstfrancis.com/hotel-features/history; "St. Francis Hotel," Clio.com, https://www.theclio.com/web/entry?id=37932.

89 **moved that afternoon to a nearby medical facility:** Merritt, *Room 1219*, 43.

89 **new diagnosis:** Ibid., 45, 63.

89 **afflicted with chronic cystitis:** Sheerin, "'Fatty' Arbuckle and Hollywood's First Scandal."

90 **"Oh, to think":** "To Think I Led Such a Quiet Life!," *Cincinnati Enquirer*, September 13, 1921.

90 **his Tudor-style mansion:** Charles F. Adams, *Murder by the Bay: Historic Homicide in and about the City of San Francisco* (Sanger, CA: Quill Driver Books/Word Dancer Press, 2005), 144.

90 **"Miss Rappe had one or two drinks":** "Probe of Death Party in S.F. Hotel Started," *Oakland Tribune*, September 10, 1921.

91 **"I am dying! I am dying!":** "Arbuckle to Be Held for Death Probe," *Oakland Tribune*, September 10, 1921.

91 **"We heard Miss Rappe moaning":** Ibid.; "Probe of Death Party in S.F. Hotel Started."

91 **"Showgirl Zey Prevon surveyed the gentlemen in the room":** "Arbuckle Witnesses in Hightower Case," *Santa Ana Register* (CA), September 30, 1921.

92 **"How long did they remain in there?":** This quote and the remainder of Prevon's statement comes from "Arbuckle Guest Gives Version of Frisco Orgy," *Arizona Republic*, September 28, 1921.

93 **"Neither I nor Mr. U'Ren nor Chief of Police":** "Arbuckle Jailed for Murder; Bail Is Denied," *Los Angeles Times*, September 11, 1921.

94 **"Police arrested him, charging him with murder":** "Brady to Ask Indictment of Film Comedian," September 12, 1921.

94 **David Bender/Arbuckle:** "Prison Mates Eager to Talk with Arbuckle," *San Francisco Chronicle*, September 12, 1921; "Arbuckle Held Without Bail," *Ogden Standard-Examiner* (UT), September 12, 1921; "Jail Doors Are Closed on Roscoe Arbuckle; Charge of Murder Follows Death of Actress; 'Now I've Got You!' Cry Ascribed to Star," *Cincinnati Enquirer*, September 12, 1921.

95 **"We ought to be friends":** "'Fatty' Arbuckle Plays Grim Real Life Role Behind Bars," *Minneapolis Star Tribune*, September 12, 1921.

95 **"I went to work this morning—incog[nito]":** Letter from Heinrich to Kaiser, September 16, 1921, box 1, John Boynton Kaiser Papers.

95 **Forensic investigation in suite:** "Microscope May Be Fateful to Fatty Arbuckle," *Salisbury Evening Post* (NC), November 14, 1921.

95 **All details from EOH's investigation in the suite:** Carton 69, folder 9–11, Edward Oscar Heinrich Papers.

95 **Collecting hair:** "Witnesses in Star's Murder Case Watched," *Los Angeles Evening Herald*, September 19, 1921.

98 **Rappe's party clothes:** Merritt, *Room 1219*, 10.

99 **"As everyone knows, I had had quite a number":** "Film Tragedy Uncovers Rum 'Road' on Coast."

99 **an underground "booze" railroad:** Ibid.

99 **"A regular system was in operation":** Ibid.

99 **"Following orders from Brady":** "Criminologist Has Evidence to Convict 'Fatty'?," *Sioux County Index* (IA), September 23, 1921.

100 **"I perceive a direct connection between":** Letter from Heinrich to Kaiser, February 14, 1921, box 1, John Boynton Kaiser Papers.

100 **"I think I lost him about two o'clock":** Letter from Heinrich to Kaiser, September 23, 1921, box 1, John Boynton Kaiser Papers.

101 **"Have made a number of important discoveries":** Letter from Heinrich to Kaiser, September 16, 1921, box 1, John Boynton Kaiser Papers.

101 **"I should think you might get a good deal of fun":** Letter from Heinrich to Kaiser, September 26, 1921, box 1, John Boynton Kaiser Papers.

101 **"You suggest that I chat with some of the maids around here":** Letter from Heinrich to Kaiser, September 23, 1921, box 1, John Boynton Kaiser Papers.

102 **"By the way the new drink down here":** Letter from Heinrich to Kaiser, September 15, 1921, box 1, John Boynton Kaiser Papers.

102 **"Fatty is guilty as hell of everything charged":** Letter from Heinrich to Kaiser, September 16, 1921, box 1, John Boynton Kaiser Papers.

102 **"Arbuckle took hold of her":** "Witness Reveals Story of 'Party,'" *Tulsa Daily World*, September 13, 1921.

102 **"The evidence adduced":** "Arbuckle Films to Be Barred," *Washington Times*, September 12, 1921.

103 **The forty-five-year-old was part prosecutor, part politician:** Scott P. Johnson, *Trials of the Century* (Santa Barbara, CA: Greenwood Publishing Group, 2010), 243.

103 **Details about coroner's jury:** "Immediate Inquest Over Body of Young Actress Is Ordered by Coroner," *Oakland Tribune,* September 12, 1921.

103 **"She said that Arbuckle threw himself on her":** "Nurse Relates to Police Story Told by Dying Actress," *Oregon Daily Journal,* September 11, 1921.

103 **"The patient admitted to me that her relations with Arbuckle":** "Witnesses Upset Case Against Arbuckle," *Daily News* (NY), September 14, 1921.

104 **"I don't like fat men":** "Immediate Inquest Over Body of Young Actress is Ordered By Coroner."

104 **"How do you know what happened if you had so many drinks":** This quote and more of Delmont's testimony from "Manslaughter Is Voted; Girls Give Evidence," *Chicago Tribune,* September 14, 1921.

105 **Grand jury deliberations:** "Evidence Held Insufficient for True Bill," *San Francisco Chronicle,* September 13, 1921.

105 **"We have sent Miss Zey Pryvon [Prevon] home under surveillance":** "State Charges Tampering with Its Witnesses," *Dayton Herald,* September 13, 1921.

105 **"I am convinced that undue influence":** Ibid.

106 **Prevon on the stand:** "One Witness Changes Story; Another Flees San Francisco," *Press and Sun-Bulletin* (Binghamton, NY), September 13, 1921.

106 **coroner's inquest:** "Brady to Make Decision on Charges Against Arbuckle," *New Castle Herald* (PA), September 15, 1921.

106 **"He was fooling with his bathrobe, kind of tying":** San Francisco Police Court, *In the Police Court of the City and County of San Francisco, State of California, Department No. 2: Honorable Sylvain J. Lazarus; The People of the State of California vs. Roscoe Arbuckle* (San Francisco: The Court, 1921), 293.

106 **"She said 'I am dying, I am dying'":** *State of California vs. Roscoe Arbuckle,* 297.

107 **"He came over and said":** Ibid., 299.

107 **"'That will make her come to'":** Ibid., 300.

107 **a Coca-Cola bottle:** Merritt, *Room 1219,* 330.

107 **"Are you aware that there is no medical evidence to back your claim":** Edmonds, *Frame-Up!,* 204.

108 **"I don't know what she was drinking":** *State of California vs. Roscoe Arbuckle,* 316.

108 **"She was plain drunk at that time, wasn't she?":** Ibid., 321.

108 **Alice Blake on the stand:** Ibid., 331–36.

108 **"I am dying, I am dying, he hurt me":** Ibid. 336.

108 **"I heard a man's voice say, 'Shut up'":** Ibid., 340.

108 **Delmont's rap sheet:** Edmonds, *Frame-Up!,* 196.

108 **Rappe's "scandalous" background:** Ibid., 213–14.

109 **"Well, I will tell you one thing, Mr. District Attorney":** *State of California vs. Roscoe Arbuckle,* 350.

109 **"We are not trying Roscoe Arbuckle alone:** David Yallop, *The Day the Laughter Stopped* (New York: St. Martin's Press, 1976).

CHAPTER 6

112 **"He put a piece of ice on her body":** "State Fires Big Gun in Hearing Yesterday Against Roscoe Arbuckle," *Morning News* (Coffeyville, KS), November 22, 1921.

112 **"Call Heinrich":** Bart Haley, "Women Witnesses Aid Arbuckle Defense in Fatty's Darkest Hour," *Evening Public Ledger* (Philadelphia, PA), November 23, 1921.

112 **"Heinrich was humorless, cold, quiet, statistical":** Ibid.

112 **McGovern sneering at him:** Oscar H. Fernbach, "Dusted Door Opens Vistas to Arbuckle," *San Francisco Examiner,* November 23, 1921.

112 **Kate Brennan testifies:** "Bar Miss Rappe's Words at Trial of Arbuckle," *Pittsburgh Post-Gazette,* November 23, 1921.

112 **All details from EOH's investigation in the suite:** Carton 69, folder 9–11, Edward Oscar Heinrich Papers.

113 **"a large amount of dust":** "The Arbuckle Trial, What Heinrich Saw Through His Microscope!," *Belfast News Letter* (Northern Ireland), November 30, 1921.

113 **Items collected:** M. D. Tracy, "Arbuckle Ready to Go on Stand," *Daily Republican* (Rushville, IN), November 25, 1921.

113 **"How do you know that among all the millions":** Haley, "Women Witnesses Aid Arbuckle Defense in Fatty's Darkest Hour."

114 **History of fingerprinting:** "History," ch. 1 in U.S. Department of Justice, *The Fingerprint Sourcebook* (Washington, D.C.: National Institute of Justice, 2013), 1–21; "Edmond Locard," The Forensics Library, http://aboutforensics.co.uk/edmond-locard.

114 **minutiae points found at the end of friction ridge:** "Fingerprint Recognition," Federal Bureau of Investigation document, https://fbi.gov/file-repository/about-us-cjis-fingerprints_biometrics-biometric-center-of-excellences-fingerprint-recognition.pdf/view.

115 **"Not all fingerprint evidence is equally good":** National Research Council, *Strengthening Forensic Science in the United States,* 8–9, 86.

116 **"She turned over on her left side":** This and all other quotes from Arbuckle's testimony from Chandler Sprague, "Arbuckle Tells Jury of Finding Girl Writhing in Agony on Bathroom Floor," *El Paso Times* (TX), November 29, 1921.

117 **"It was the matter of fingerprints purely in the final analysis":** "Arbuckle Woman Juror Charges Intimidations," *Oakland Tribune,* December 5, 1921.

117 **"The ability of the defense to create":** Letter from Heinrich to Kaiser, December 3, 1921, box 1, John Boynton Kaiser Papers.

118 **"Parents seem to think they must watch their sons":** Ibid.

118 **"The case of Arbuckle and its present status":** Ibid.

118 **Goals of Prohibition:** William E. Nelson, "Criminality and Sexual Morality in New York, 1920–1980," *Yale Journal of Law & the Humanities* 5, no. 2 (May 2013): 269.

118 **often young women were cruelly chastised in court:** Estelle B. Freedman, *Redefining Rape: Sexual Violence in the Era of Suffrage and Segregation* (Cambridge, MA: Harvard University Press, 2013), 147–48, 160, 191.

118 **"Our nation depends for its existence":** Letter from Heinrich to Kaiser, December 3, 1921, box 1, John Boynton Kaiser Papers.

119 **doomed king Belshazzar in Babylon:** "Belshazzar's Party," Daniel 5:1–31 (Common English Bible).

119 **"Like Daniel in the days of Babylon":** Letter from Heinrich to Kaiser, December 3, 1921, box 1, John Boynton Kaiser Papers.

119 **Christmas in 1921 was glorious for many Americans:** Angela Meiquan Wang, "A Christmas Wish List in the 1920s," *BuzzFeed,* November 28, 2012.

119 **Christmas dinner menu:** *Good Housekeeping's Book of Menus, Recipes and Household Discoveries* (New York: Good Housekeeping, 1922), 49–50.

120 **Salvation Army history:** "History of the Salvation Army," The Salvation Army, https://www.salvationarmyusa.org/usn/history-of-the-salvation-army.

120 **"When Santa Claus came to me instead":** Letter from Heinrich to Kaiser, December 20, 1921, box 1, John Boynton Kaiser Papers.

121 **"If I hurry Marion too much":** Letter from Heinrich to Kaiser, September 15, 1921, box 1, John Boynton Kaiser Papers.

121 **"It has been my privilege and pleasure":** Letter from Heinrich to Kaiser, May 10, 1921, box 1, John Boynton Kaiser Papers.

122 **"I am not positive that I am doing yet":** "Edward Oscar Heinrich," *California Monthly* (Berkeley, CA: Cal Alumni Association, February 1926), 344.

122 **"The right hand of the man clutching the hand":** "Fingerprint Sharp in Arbuckle Case," *Reno Gazette-Journal,* January 23, 1922.

123 **"Is it not a fact that you introduced yourself":** "Former Suitor of Miss Rappe Aids Arbuckle," *Oakland Tribune,* November 26, 1921.

123 **"in order not to be disturbed":** "Fingerprint Sharp in Arbuckle Case."

124 **"Permit me to say that this remittance":** Letter from Heinrich to E. O. Tisch, April 21, 1921, box 30, folder 26, Edward Oscar Heinrich Papers.

124 **"Please don't feel backward about getting wood":** Letter from Heinrich to his mother, October 22, 1921, box 29, folder 40–41, Edward Oscar Heinrich Papers.

124 **"Don't be worried about anybody you":** Ibid.

125 **"Can a spirit humbled by adversity be pompous":** Letter from Heinrich to Kaiser, October 9, 1922, box 1, John Boynton Kaiser Papers.

125 **"In childhood and youth":** Ibid.

125 **"Mr. Arbuckle came within an ace of being convicted":** Letter from Heinrich to Charles Hardless Jr., Esq., July 6, 1922, box 27, folder 18, Edward Oscar Heinrich Papers.

126 **"One ballot, no talk":** "Arbuckle Hopes to Do a Comeback," *Boston Globe,* April 13, 1922.

126 **"If the public doesn't want me, then I'll take my medicine":** Ibid.

127 **"I never had a cake or even a reminder":** Letter from Heinrich to Kaiser, April 21, 1921, box 1, Edward Oscar Heinrich Papers.

127 **"I move more deliberately, I require more sleep":** Ibid.

127 **"Did I tell you that Theodore could sing the song":** Letter from Heinrich to his mother, November 25, 1921, box 29, folder 40–41, Edward Oscar Heinrich Papers.

127 **"If the entire episode results":** Letter from Heinrich to Kaiser, December 3, 1921, box 1, John Boynton Kaiser Papers.

127 **Minta Durfee, divorced him:** "Wife of Fatty Arbuckle Gets Divorce in Paris," *Chicago Tribune,* January 27, 1925.

128 **married his mistress:** "Arbuckle Hit as Sheik of Beach Party," *Daily News* (NY), August 6, 1928.

128 **Los Angeles mansion:** "Arbuckle Has $100,00 Home," *San Francisco Chronicle,* September 12, 1921.

128 **"It was a dismal experience to watch him":** Buster Keaton, *My Wonderful World of Slapstick* (New South Wales, Australia: Allen & Unwin, June 1967).

128 **Heinrich inspired Arbuckle:** Yallop, *The Day the Laughter Stopped,* 278.

CHAPTER 7

129 **"We will play it on one card":** This quote and much of the recollection of Roy DeAutremont comes from *Trial Transcript of Roy DeAutremont on the Train Robbery in Siskiyou Tunnel,* Ashland Library, "Oregon Cabinet," Call # ORE CAB 364.1552 TRI, 18.

129 **The recollections of Hugh and Ray:** "Confession of Hugh DeAutremont," June 23, 1927, Oregon State Archives DeAutremont Collection, Eugene, OR; "Confession of Ray DeAutremont," June 23, 1927, Oregon State Archives DeAutremont Collection, Eugene, OR.

129 **Description of Siskiyou County:** Pepper Trail and Edgard Espinoza, "Tunnel 13: How Forensic Science Helped Solve America's Last Great Train Robbery," Jefferson Public Radio, December 31, 2013.

129 **Location of summit:** "Four Hold up Men Blow up Mail Car and Make Escape," *Statesman Journal* (Salem, OR), October 12, 1923.

130 **steepest railway in the country:** *Murder on the Southern Pacific,* Oregon Historical Society, https://www.opb.org/artsandlife/series/historical-photo/oregon-historical-photo-chinese-rail-workers.

130 **The Road of a Thousand Wonders:** Southern Pacific Company, *The Road of a Thousand Wonders: The Coast Line–Shasta Route of the Southern Pacific Company from Los Angeles Through San Francisco, to Portland, a Journey of Over One Thousand Three Hundred Miles* (San Francisco: Southern Pacific Co., 1908).

130 **Gold Special:** "D'Autremonts' 1923 Escapade Marks Its 75th Anniversary," *Santa Maria Times* (CA), October 12, 1998.

130 **Number of cars:** Scott Mangold, *Tragedy at Southern Oregon Tunnel 13: DeAutremonts Hold Up the Southern Pacific* (Charleston: The History Press, 2013), 25; Alan Hynd, "The Case of the Murders in Tunnel 13," *Time,* May 12, 1930.

131 **A "blasting machine," a small wooden red box:** Details about the DuPont detonator from Heinrich notes on a radio program, carton 28, folder 34, Edward Oscar Heinrich Papers.

131 **Blasting machine:** "What Is this Mining Device?," Cave Creek Museum fact sheet.

131 **DuPont box stolen:** Nancy Pope, "DeAutremont Brothers Train Robbery," *Pushing the Envelope,* October 11, 2012, http://postalmuseumblog.si.edu/2012/10/deautremont-brothers-train-robbery.html.

131 **"Roy carried a 45-caliber Colt":** The photo of the .45-caliber gun from Heinrich's file on train robbery, carton 70, folder 15, oversize box 1, folder 3–4, Edward Oscar Heinrich Papers.

131 **Brothers' history with guns:** *Trial Transcript of Roy DeAutremont,* 3.

131 **"What do you think about it, little lad?":** Ibid., 19.

131 **$40,000 worth of gold:** Pope, "DeAutremont Brothers Train Robbery."

132 **In the annals of great American Western films:** Description of *Great Train Robbery* comes from the short film directed by Edwin S. Porter in 1903, along with its IMDb entry.

132 **The role of movies in real robberies:** "Murder on the Southern Pacific," *Oregon Experience,* season 9, episode 904, Oregon Public Broadcasting, May 3, 2015.

132 **DeAutremonts loved Westerns:** Audio interview with Ray DeAutremont by Gary Williams, tape FT-12658, 1973, Ed Kahn Collection, University of North Carolina, Chapel Hill.

133 **Marines guarding U.S. Mail:** George Corney, "Crime and Postal History: Bring in the Marines!," Marine Corps Association & Foundation, 1993, https://www.imdb.com/title/tt0000439/.

133 **Vice President Calvin Coolidge took office:** "Calvin Coolidge Biography," Biography.com, April 2, 2014, last updated April 17, 2019.

133 **After Harding's death:** "Warren G. Harding," *Encyclopaedia Britannica,* July 3, 2019.

133 **quickly grew by 7 percent:** "The Business of America: The Economy in the 1920s," Roaring Twenties Reference Library, Encyclopedia.com, 2006.

133 **"Hugh, you see what is in front of you":** *Trial Transcript of Roy DeAutremont,* 19.

134 **Background on the brothers' early lives:** Ibid., various pages.

134 **"Their marriage life got to be worse and worse":** Ibid., 4.

135 **History of the Wobblies:** "Murder on the Southern Pacific."

135 **criminal syndicalism:** "Criminal Syndicalism Laws," *Encyclopaedia Britannica,* July 20, 1998.

135 **"I did not know him, he was so changed":** *Trial Transcript of Roy DeAutremont,* 8.

135 **Ray's plan in prison:** "Confession of Ray DeAutremont," DeAutremont Collection, 1.

135 **"We knew it would mean we couldn't help mother":** *Trial Transcript of Roy DeAutremont,* 18.

136 **But they were also slim and short:** Details about height and weights from the Oregon State Penitentiary, Salem, Oregon, June 24, 1927, carton 5, Southern Pacific Company Train Robbery Records, 1892–1940, BANC CA-372, Bancroft Library, University of California, Berkeley.

136 **more than sixty spent .45-caliber shells:** Letter from a special agent with Southern Pacific Company to C. E. Terrill, the sheriff of Jackson County, November 10, 1923, Southern Oregon Historical Society Collection.

136 **Shooting practice:** *Trial Transcript of Roy DeAutremont,* 17.

136 **spending weeks traversing the countryside:** "Confession of Ray DeAutremont," DeAutremont Collection,

137 **Federal agent and Hugh:** *Trial Transcript of Roy DeAutremont,* 18.

138 **"I hit the platform with my knee":** Ibid.

138 **"Boys, I don't give a damn":** Ibid., 19.

138 **"It was pretty near always on time":** Ibid.

139 **"Hugh, go out":** Ibid.

139 **Confusion over pistols:** In Ray's testimony, he says Roy used his .45-caliber to shoot the fireman, but Roy's testimony confirms that he dropped it where deputies found a .45-caliber.

139 **"The worst I was scared on the whole job":** *Trial Transcript of Roy DeAutremont,* 19.

140 **mail clerk Elvyn Dougherty:** "D'Autremonts' 1923 Escapade Marks Its 75th Anniversary."

140 **There was a tremendous explosion:** Details about the placement of the dynamite have varied in different secondary books about this case, but Heinrich confirmed that the TNT was laid at the mouth of the tunnel in his notes on a radio program, carton 28, folder 34, Edward Oscar Heinrich Papers.

140 **overturning a coal-burning stove:** Notes on a radio program, carton 28, folder 34, Edward Oscar Heinrich Papers; and "Four Hold Up Men Blow Up Mail Car and Make Escape."

140 **"I killed the mail clerk":** *Trial Transcript of Roy DeAutremont,* 20.

140 **glass shattered around the passengers:** "Four Hold Up Men Blow up Mail Car and Make Escape."

141 **"I told him that his life was in greater danger":** *Trial Transcript of Roy DeAutremont,* 20.

141 **"That other fellow said":** "Confession of Ray DeAutremont," DeAutremont Collection, 2.

142 **"I shot him with his hands in the air":** *Trial Transcript of Roy DeAutremont,* 20.

142 **only Sid Bates, the engineer, was left alive:** Mangold, *Tragedy at Southern Oregon Tunnel 13,* 13.

144 **Oscar Heinrich lingered over the collection of evidence:** Work journal, October 16, 1923, folder labeled "Siskiyou train robbery," carton 70, folder 15, oversize box 1, folder 3–4, Edward Oscar Heinrich Papers, 1.

144 **baggage clerk saw just two men:** "Train Bandits Kill Four Men in California," *Baltimore Sun*, October 12, 1923.

144 **"Nothing doing":** "Bandit Suspect in California," *Medford Mail Tribune* (OR), October 16, 1923.

144 **the murder of Anna Wilkens:** "Wilkens Is Arrested, Charged with Murder," *Madera Mercury* (CA), July 15, 1922.

145 **Hollywood director and actor William Desmond Taylor:** Christopher Hudspeth, Ryan Bergara, and Shane Madej, "The Murder of William Desmond Taylor Is One of the Most Peculiar Unsolved Mysteries," *BuzzFeed,* March 2, 2018.

145 **"The more learned the chemist":** Letter from Heinrich to Kaiser, October 9, 1922, box 12, folder 27, Edward Oscar Heinrich Papers.

145 **Information on the search:** "4 Trainmen Slain in Holdup, Posses Search Mountains," *St. Louis Star and Times,* October 11, 1923; "Mail Bandit Gang Faces Lynching," *Santa Ana Register* (CA), October 12, 1923; "Mail Car Dynamited by Bandits in Daring Raid; Loot Unknown," *Indianapolis Star,* October 12, 1923; "Bloodhounds and Posses Trailing Bandits Who Robbed Train and Murdered Crew," *Roseburg News-Review,* (OR), October 12, 1923.

145 **"Bloodhounds today failed to pick up the scent of the desperadoes":** "S.P. Offers $2500 for Bandits," *Woodland Daily Democrat* (CA), October 13, 1923.

146 **already talk of lynching:** "Big Manhunt in Mountains of Western States," *Mansfield News* (OH), October 12, 1923.

146 **"It was the boldest train robbery since":** Ibid.

146 **Police briefly arrested a twenty-two-year-old ex-convict:** "Man Arrested in Train Holdup," *Des Moines Register,* October 14, 1923.

146 **They held two drug addicts:** "Bandit Suspect in California."

146 **Three hunters admitted:** "Prisoners Questioned by Officers," *La Grande Observer* (OR), October 18, 1923.

147 **Deputies interview mechanic:** Block, *The Wizard of Berkeley,* 14–15.

147 **Agents had searched the overalls:** Notes on a radio program, carton 28, folder 34, "Siskiyou train robbery," Edward Oscar Heinrich Papers, 1.

147 **"He held up the overalls to a wooden door":** Details about the overalls tacked on a door from photographs found in carton 70, folder 15, oversize box 1, folder 3–4, Edward Oscar Heinrich Papers. Another discrepancy in secondary sources has been that the legs of the overalls had been tucked inside the boots, which Heinrich dispelled in his notes on a radio program, carton 28, folder 34, "Siskiyou train robbery," Edward Oscar Heinrich Papers, 9.

148 **Heinrich's logging shoes:** Ibid., 8–9.

148 **"The overalls were quite new":** Ibid., 9.

148 **"Suspenders on left fastened and unfastened habitually":** Note found in carton 70, folder 15, oversize box 1, folder 3–4, undated, Edward Oscar Heinrich Papers.

148 **"Every individual, particularly a man, accumulates dust":** Letter from Heinrich to Professor F. W. Martin, March 15, 1924, carton 85, folder 189, Edward Oscar Heinrich Papers.

149 **Components of grease:** "Grease—Its Components and Characteristics," Exxon Mobil Corporation, 2009.

149 **the same type of naturally occurring, sticky resin:** "Calking," Traditional Maritime Skills, Maritime Heritage Skills Cluster Project, http://www.boat-building.org/learn-skills/index.php/en/wood/caulking-calking.

149 **"No larger than the size of half a pea":** Notes on a radio program, carton 28, folder 34, "Siskiyou train robbery," Edward Oscar Heinrich Papers, 10.

149 **"The pockets carried tiny chips":** Letter from Heinrich to Professor F. W. Martin, March 15, 1924, carton 85, folder 189, Edward Oscar Heinrich Papers.

149 **Oscar used a small electric suction:** Hynd, "The Case of the Murders in Tunnel 13."
149 **"A man who carries a fingernail file":** Ibid.
150 **"Pencil pocket at the left side of the bib":** Heinrich's affidavit, carton 70, folder 15, oversize box 1, folder 3–4, "Siskiyou train robbery," Edward Oscar Heinrich Papers.
150 **"I aim to return evidence as intact":** Block, *The Wizard of Berkeley,* 20.
150 **"about the size of a cigarette paper":** Notes on a radio program, carton 28, folder 34, "Siskiyou train robbery," Edward Oscar Heinrich Papers, 11.
150 **Next Oscar examined the Colt .45-caliber revolver:** Details from the gun come from scraps of paper found in carton 70, folder 15, oversize box 1, folder 3–4, Edward Oscar Heinrich Papers.
150 **firearms manufacturers had inscribed a second set:** Jay Robert Nash, *Bloodletters and Badmen: A Narrative Encyclopedia of American Criminals from the Pilgrims to the Present* (New York: M. Evans and Company, 1973), 151.
151 **"The wearer and owner was a lumberjack":** Part of a twenty-six-page sworn affidavit for Dan O'Connell, carton 70, folder 15, oversize box 1, folder 3–4, Edward Oscar Heinrich Papers.
151 **"Not so fast, professor":** Hynd, "The Case of the Murders in Tunnel 13."
151 **"Hence the owner":** Notes on a radio program, carton 28, folder 34, "Siskiyou train robbery," Edward Oscar Heinrich Papers, 8–9.
151 **that didn't necessarily mean that the man was left-handed:** From the lore of this case comes a common belief that Heinrich stated that the owner of the overalls was left-handed, but I can find no proof in his own comprehensive files that he came to that conclusion.
151 **Validity of hair analysis:** National Research Council, *Strengthening Forensic Science in the United States*, 121.
152 **"Putting these things together":** Letter from Heinrich to Professor F. W. Martin, March 15, 1924, carton 85, folder 189, Edward Oscar Heinrich Papers.
152 **"*The Black Kit Bag. By C. O. Heinrich*" was neatly printed":** *The Black Kit Bag* comes from a manuscript found in carton 70, folder 15, oversize box 1, folder 3–4, November 2, 1923, Edward Oscar Heinrich Papers.
153 **"I am interested in working out those defects":** Letter from Heinrich to his mother, May 8, 1924, box 29, folder 40–41, Edward Oscar Heinrich Papers.
153 **"Your special interest in babies, weeds, legal aid societies":** Letter from Kaiser to Heinrich, September 23, 1918, box 12, folder 27, Edward Oscar Heinrich Papers.
154 **"I have two, and possibly three novels":** Letter from Heinrich to Kaiser, October 9, 1922, box 12, folder 27, Edward Oscar Heinrich Papers.
154 **"I am not wishing to write to win fame or fortune":** Ibid.
154 **Carfare on Thursday had cost him twelve cents:** Details about Heinrich's ledgers from come from scraps of paper in carton 70, folder 15, oversize box 1, folder 3–4, November 2, 1923, Edward Oscar Heinrich Papers.
155 **He stored evidence from many investigations:** Heinrich's ephemera from cases can be found in carton 70, folder 15, oversize box 1, folder 3–4, November 2, 1923, Edward Oscar Heinrich Papers.

155 **"The paint may be applied to dry walls":** Letter from Heinrich to his mother, August 22, 1923, box 29, folder 40–41, Edward Oscar Heinrich Papers.

156 **charted his own urine levels:** These come from a chart in Edward Oscar Heinrich Papers; a letter from Heinrich to Theodore, August 20, 1940, box 23, 89–44, file 182, Theodore Heinrich Collection.

156 **"Of course we are glad you are busy":** Letter from Marion to Heinrich, August 3, 1924, box 27, Edward Oscar Heinrich Papers.

156 **"I hope now that conditions will adjust themselves":** Letter from Heinrich to his mother, May 8, 1924, box 29, folder 40–41, Edward Oscar Heinrich Papers.

156 **"My work has been very light and expense very heavy":** Letter from Heinrich to his mother, November 23, 1922, box 29, folder 40–41, Edward Oscar Heinrich Papers.

157 **"His determination to stick to the face value":** Letter from Heinrich to J. F. Dennisen, September 22, 1923, box 5, folder 53, Edward Oscar Heinrich Papers.

157 **a dingy shack that could scarcely be called a cabin:** Details from the cabin and Camp No. 2 come from photos and their captions in Southern Pacific Company train robbery records (1892–1940), BANC MSS C-A 372, carton 5 and carton 70, folder 15, oversize box 1, folder 3–4, Edward Oscar Heinrich Papers.

157 **The U. S. Postal Service paper:** Colin Wilson, *Written in Blood* (New York: Diversion Books, 2015), 314.

158 **Oscar compared one hair on the towel in the cabin:** Details about Heinrich's conclusion come from various pages from carton 70, folder 15, oversize box 1, folder 3–4, Edward Oscar Heinrich Papers.

158 **"The manhunt was astonishing":** Trail and Espinoza, "Tunnel 13"; "Real Sherlock Holmes, with Four Slender Clews, Pulled Net Around DeAutremonts," *News-Herald* (OH), June 27, 1927.

159 **"There is no trick at all to visualize":** Letter from Heinrich to Professor F. W. Martin, March 15, 1924, carton 85, folder 189, Edward Oscar Heinrich Papers.

159 **"I feel a strong desire to sit:** Letter from Heinrich to his mother, May 8, 1924, box 29, folder 40–41, Edward Oscar Heinrich Papers.

159 **"About this story":** Letter from editor to Heinrich, September 12, 1924, box 12, folder 27, Edward Oscar Heinrich Papers.

160 **"The last person in the world one would":** "Real Sherlock Holmes, with Four Slender Clews, Pulled Net Around DeAutremonts."

CHAPTER 8

164 **Schwartz age and background:** Bosworth and Jackson, eds., *San Francisco Murders*, 247.

164 **"The Pacific Cellulose company":** From an article on page 38 of *Oakland Tribune*, June 14, 1925.

164 **silkworms:** "Silkworm Moth," *Encyclopaedia Britannica*, September 29, 2006.

164 **Price of silk dress:** "Historic Prices—1927," Morris County Library, https://mclib.info/reference/local-history-genealogy/historic-prices/1927-2.

165 **"heart balm" lawsuit:** Tori Telfer, "How the 'Heart Balm Racket' Convinced America That Women Were Up to No Good," *Smithsonian Magazine,* February 13, 2018.

165 **Vollmer is sued:** "Wife Stands by Vollmer," *Los Angeles Times,* August 8, 1924.

166 **"The case is far too serious to use spectacular":** "Former Wife of Sued Chief May Assist Woman," *Oakland Tribune,* August 9, 1924.

166 **"Please reserve judgment on this for some time":** Letter from Heinrich to his mother, August 12, 1924, box 29, folder 40–41, Edward Oscar Heinrich Papers.

166 **"This is merely a plot to discredit me in my business":** "Girl Sues Wedded Man for $75,000," *Oakland Tribune,* June 9, 1925.

166 **"Someday, my dear":** "When Justice Triumphed," *Daily News* (NY), June 30, 1929.

166 **chemist Charles Schwartz denied wooing the young woman:** Details about July 30 from various newspaper articles and in Bosworth and Jackson, eds., *San Francisco Murders,* 255.

167 **Clear liquid sloshed from a vat:** Details about liquids inside lab from Heinrich's typed memo from work journals in case file, carton 74, folder 14, Edward Oscar Heinrich Papers.

167 **Night watchman's recollections:** Heinrich's typed memo from ibid., 10.

167 **"I plan to do some experimenting with ether":** "When Justice Triumphed," *Daily News* (NY), June 30, 1929.

168 **"There was music from my neighbor's house":** F. Scott Fitzgerald, *The Great Gatsby* (New York: Charles Scribner's Sons, 1925), 39.

168 **the Roaring Twenties:** "The Roaring Twenties History," History.com, April 14, 2010, https://www.history.com/topics/roaring-twenties/roaring-twenties-history; "The Roaring Twenties: 1920–1929," Boundless US History, *Lumen Candela,* https://courses.lumenlearning.com/boundless-ushistory/chapter/the-roaring-twenties.

169 **"I have faith in my invention":** "When Justice Triumphed."

170 **"I will send you something extra a little later as my money comes in":** Letter from Heinrich to his mother, April 24, 1925, box 29, folder 40–41, Edward Oscar Heinrich Papers.

170 **"The funeral expenses of dead horses":** Letter from Heinrich to his mother, May 8, 1924, box 29, folder 40–41, Edward Oscar Heinrich Papers.

170 **"My professional calls require me to travel":** Letter from Heinrich to Bennett F. Davenport, November 18, 1924, box 26, folder 12, Edward Oscar Heinrich Papers.

171 **"I have an idea that I could write":** "The Detective and the Chemist," *St. Louis Post-Dispatch,* November 9, 1924.

171 **The room was spacious but very bright:** Details about undertaker's room come from Heinrich's typed memo in case file, carton 74, folder 14, Edward Oscar Heinrich Papers.

172 **"Exposures were made in each case":** Ibid., 1.

173 **"He called me on the telephone":** Block, *The Wizard of Berkeley,* 114.

173 **"I've seen the body and it's that of Mr. Schwartz":** Ibid., 116.

174 **"He was murdered by people":** Ibid.
174 **no gas or water sources:** "Quick Finish of Slayer Expected by Crime Expert," *Oakland Tribune,* August 10, 1925.
175 **"Five gallon can of carbon disulfide":** Heinrich's typed memo (p. 6) in case file, carton 74, folder 14, Edward Oscar Heinrich Papers.
175 **"A Sikes office chair was in the path of the combustion":** Ibid., 9.
175 **"Did the explosion start":** "Unraveling of Crime Skein Detailed by Criminologist," *San Francisco Examiner,* August 10, 1925.
175 **"Cracks in flooring on east":** Heinrich's typed memo (pp. 10–11) in case file, carton 74, folder 14, Edward Oscar Heinrich Papers.
176 **"By tracing the flames on the floor":** Block, *The Wizard of Berkeley,* 121.
176 **"Projection group—consists of several large stains":** Heinrich's typed memo (p. 5) in case file, carton 74, folder 14, Edward Oscar Heinrich Papers.
176 **it had dripped through the closet floor:** Ibid., 13.
176 **"That's the tooth I extracted not very long ago":** Ibid., 117.
176 **"Replaced the eye in socket prior to making photographs":** Heinrich's typed memo (p.4) in case file, carton 74, folder 14, Edward Oscar Heinrich Papers.
176 **When Oscar later dissected the eyeball:** Heinrich's work journal, August 7, 1925, carton 74, folder 14, Edward Oscar Heinrich Papers.
176 **"Every photograph of my husband":** Block, *The Wizard of Berkeley,* 117.
177 **his watch and change:** Block, *The Wizard of Berkeley,* 115.
178 **Schwartz's height:** Bosworth and Jackson, eds., *San Francisco Murders,* 263.
178 **"I'd like to know, first of all":** This quote and all details of Heinrich's findings from Block, *The Wizard of Berkeley,* 117–21.
179 **"He tried to give the impression":** "Missing Chemist in Torso Mystery 'Perfect Crimes' Student, Police Declare," *San Bernardino County Sun,* August 24, 1925.
180 **"What interests me about murderers, captain":** "When Justice Triumphed."
180 ***The Philosophy of Eternal Brotherhood.* It had been found on the body:** Found in the "Schwartz" file, August 7, 1925, carton 74, folder 14, Edward Oscar Heinrich Papers.
181 **an advertisement in a San Francisco newspaper:** Bosworth and Jackson, eds., *San Francisco Murders,* 253.
181 **Details of the bindle:** "Quick Finish of Slayer Expected by Crime Expert."
181 **an undertaker in Placerville, California, told police:** Block, *The Wizard of Berkeley,* 126.
182 **Schwartz's deceptions:** Bosworth and Jackson, eds., *San Francisco Murders,* 247–54; Block, *The Wizard of Berkeley,* 117.
182 **"Open that door—police!":** Ibid., 127.
183 **Poison tablets were beside his bed:** "Schwartz's Suicide Bares Deliberate Murder Plan," *San Francisco Examiner,* August 10, 1925.
183 **"The only thing I did was I tried to burn him":** "When Justice Triumphed."
183 **"Schwartz was too familiar with crime detection":** "Murder Plot Being Probed," *San Bernardino County Sun,* August 7, 1925.

183 **"It was too perfect to have been":** Block, *The Wizard of Berkeley,* 122.

184 **"The criminal who leaves room for a chance":** "When Justice Triumphed."

184 **"some expensive mistakes [that] had been made in the past":** Letter from Heinrich to his mother, May 8, 1924, box 29, folder 40–41, Edward Oscar Heinrich Papers.

CHAPTER 9

185 **The saltwater marshes in El Cerrito:** "Girl's Death Indicated in Mystery Find," *San Francisco Examiner,* August 24, 1925.

186 **"New murder case":** Heinrich's work journal, August 24, 1925, carton 70, folder 36–37, Edward Oscar Heinrich Papers.

186 **A dainty, delicate ear slid:** Details about examination results in letter from Heinrich to Earl Warren titled "In re: Bessie Ferguson," June 24, 1926, "Ferguson" folder, carton 70, folder 36–37, Edward Oscar Heinrich Papers.

187 **Detectives entered every duck blind and hunting lodge:** "Nurse Was Killed Outright Is Now Thought by Officers," *Reno Gazette-Journal,* August 28, 1925.

187 **In his lab Oscar picked up a hatchet:** "Girl Murder Revealed as Ear and Scalp Are Found by Boy in Richmond Marsh," *Oakland Tribune,* August 24, 1925.

187 **"I do not know how blood stains":** "Girl Murder Revealed as Ear and Scalp Are Found by Boy in Richmond Marsh."

188 **The study of insects in crime solving:** "Investigating Forensics," Simon Fraser University Museum of Archaeology and Ethnology, 2010; Y. Z. Erzincllioğlu, "The Application of Entomology to Forensic Medicine," *Medicine, Science and the Law* 23, no. 1 (1983): 57–63, http://www.sfu.museum/forensics/eng/.

189 **"Assuming an additional twenty-four hours":** Letter from Heinrich to Earl Warren titled "In re: Bessie Ferguson," June 24, 1926, "Ferguson" folder, carton 70, folder 36–37, Edward Oscar Heinrich Papers.

189 **"Small fragments of plaster, coal, decayed redwood":** Ibid., 6.

190 **"The city editor is, without exception":** Letter from Heinrich to Kaiser, December 26, 1926, box 1, John Boynton Kaiser Papers.

191 **"Size of rock particles indicate alternate current":** Heinrich notes, August 30, 1925, "Ferguson" folder, carton 70, folder 36–37, Edward Oscar Heinrich Papers.

191 **"It's somewhere around Bay Farm Island":** This quote and the remainder of conversation from Block, *The Wizard of Berkeley,* 138–40.

192 **quartz grain surface textures:** Alastair Ruffell and Jennifer McKinley, "Spatial Distribution of Soil Geochemistry in Geoforensics," *Unearthed: Impacts of the Tellus Surveys of the North of Ireland,* ed. Mike Young (Dublin: Royal Irish Academy, 2016).

192 **atomic force microscopes:** D. Konopinski et al., "Investigation of Quartz Grain Surface Textures by Atomic Force Microscopy for Forensic Analysis," *Forensic Science International* 22, no. 1–3 (November 2012): 245–55; "Atomic Force Microscopes," an information sheet published by the Bruker Corporation.

192 **"He is well balanced, quiet, courteous, and helpful":** Letter from Heinrich to his mother, May 8, 1924, box 29, folder 40–41, Edward Oscar Heinrich Papers.

193 **"My dear big boy":** Letter from Heinrich to Theodore, June 13, 1925, 89–44, box 23, file 178, Theodore Heinrich Collection.

193 **"The book which you remember":** Letter from Heinrich to Bennett F. Davenport, October 24, 1924, box 26, folder 12, Edward Oscar Heinrich Papers.

194 **"Examined a skull which has been cut in several":** Letter from Heinrich to Earl Warren titled "In re: Bessie Ferguson," June 24, 1926, "Ferguson" folder, carton 70, folder 36–37, Edward Oscar Heinrich Papers.

194 **flesh using chemicals:** "Heinrich Turns Over Tule Clues to Authorities," *San Francisco Examiner*, September 18, 1925.

194 **"It may be tentatively assumed that dismemberment":** Letter from Heinrich to Earl Warren titled "In re: Bessie Ferguson" (p. 6), June 24, 1926, "Ferguson" folder, carton 70, folder 36–37, Edward Oscar Heinrich Papers.

194 **an Oakland dentist confirmed:** "Slain Girl Identified as Oakland Nurse by Her Family and Dentist," *Oakland Tribune*, August 25, 1925.

195 **He asked about their family background:** Details about Bessie's family life come from Heinrich notes, September 28, 1925, "Ferguson" folder, carton 70, folder 36–37, Edward Oscar Heinrich Papers.

196 **Bessie Ferguson always had spending money:** "Nurse Was Killed Outright Is Now Thought by Officers."

196 **expensive bags, stylish clothes, and even a diamond ring:** "Eastbay Business Men Linked with Tule Death," *Oakland Tribune*, August 27, 1925.

196 **"The only way is to have her away":** "Letters to Aid Authorities," *San Francisco Chronicle*, August 29, 1925.

196 **"We never know what you are going to do":** Ibid.

197 **"I won't be with him long, though":** Block, *The Wizard of Berkeley*, 132–33.

197 **"The work is neatly done":** Letter from Heinrich to Earl Warren titled "In re: Bessie Ferguson," "Ferguson" folder, carton 70, folder 36–37, Edward Oscar Heinrich Papers.

197 **Jack the Ripper:** Scott A. Bonn, "Jack the Ripper Identified," *Psychology Today*, January 27, 2014.

197 **Serial killers and mass murderers:** Scott A. Bonn, "Origin of the Term 'Serial Killer,'" *Psychology Today*, June 9, 2014.

198 **"The skill shown requires the candidate:** Letter from Heinrich to Earl Warren titled "In re: Bessie Ferguson," June 24, 1926, "Ferguson" folder, carton 70, folder 36–37, Edward Oscar Heinrich Papers.

198 **The killer likely had thick, dark brown hair:** "Trace Killer by Two Dark Hairs," *Bakersfield Californian*, September 16, 1925.

199 **"She tried to persuade me by her deportment":** Letter from Heinrich to his mother, July 10, 1924, box 29, folder 40–41, Edward Oscar Heinrich Papers.

199 **"Marion has been the lure":** Letter from Heinrich to Kaiser, May 7, 1946, box 1, John Boynton Kaiser Papers.

199 **"Marion moves along in an orderly way":** Letter from Heinrich to his mother, May 8, 1924, box 29, folder 40–41, Edward Oscar Heinrich Papers.

199 **Oscar could smell death:** Details regarding Heinrich's visit to hunting lodge come from a letter from Heinrich to Earl Warren titled "In re: Bessie Ferguson" (p. 12), June 24, 1926, "Ferguson" folder," carton 4, folder 12–14, Edward Oscar Heinrich Papers.

200 **"Says Barnett wanted her to go to a doctor":** Notes, September 28, 1925, "Ferguson" folder, carton 4, folder 12–14, Edward Oscar Heinrich Papers.

200 **"The skill shown in the dismemberment":** Letter from Heinrich to Earl Warren titled "In re: Bessie Ferguson," June 24, 1926, "Ferguson" folder, carton 12, folder 6–7, Edward Oscar Heinrich Papers.

201 **"Talking will fatigue men":** Lecture notes, July 19, 1920, "Criminology 113C: course materials and lectures 1920" folder, carton 12, folder 6–7, Edward Oscar Heinrich Papers.

201 **The students stared back at their professor:** "Sherlock Holmeses in Embryo Should Attend This University Course," *Santa Cruz Evening News* (CA), February 24, 1917.

202 **"A little knowledge is a dangerous thing":** Letter from Kaiser to Heinrich, July 5, 1921, box 1, John Boynton Kaiser Papers.

202 **"Typical of the policeman who has been sold in scientific investigation":** Letter from Heinrich to Kaiser, May 16, 1927, box 28, folder 14, Edward Oscar Heinrich Papers.

202 **"The moral seems to be that":** Letter from Heinrich to Crossman, October 18, 1926, box 5, folder 30, Edward Oscar Heinrich Papers.

202 **"I have long held the opinion":** Letter from Heinrich to Bennett F. Davenport, November 18, 1924, box 26, folder 12, Edward Oscar Heinrich Papers.

203 **"I do not yet know whether to look upon":** Letter from Heinrich to Bennett F. Davenport, July 23, 1925, box 26, folder 12, Edward Oscar Heinrich Papers.

203 **"I found impressed in the fibres":** Letter from Heinrich to Earl Warren titled "In re: Bessie Ferguson," June 24, 1926, "Ferguson" folder, carton 12, folder 6–7, Edward Oscar Heinrich Papers.

204 **Frank Barnet:** "Perspectives of a Newspaperwoman," *Perspectives on the Alameda County District Attorney's Office*, Vol. I. (1970).

205 **"I am not prepared to say":** Letter from Heinrich to Earl Warren titled "In re: Bessie Ferguson," June 24, 1926, "Ferguson" folder, carton 12, folder 6–7, Edward Oscar Heinrich Papers.

205 **A hiker traversing a trail in El Cerrito:** "Gruesome Find Starts Probe of Mystery Pit Full of Charred Bones," *Santa Cruz Evening News* (CA), November 2, 1927.

CHAPTER 10

207 **Martin Colwell stood at the door:** Colin Evans, *Murder 2: The Second Casebook of Forensic Detection* (New York: John Wiley & Sons, 2004), 60; Block, *The Wizard of Berkeley*, 166–67.

209 **"Wall bullet: 144.69 grams":** Notes on experiments, "Colwell" folder, carton 70, folder 1–3, Edward Oscar Heinrich Papers.

209 **History of ballistics:** Thomas Gale, "Microscope, Comparison," *World of Forensic Science* (Detroit: Thomson Gale, 2006); Lisa Steele, *Science for Lawyers* (Chicago: American Bar Association, 2008), 1–4.

209 **The silver object came into focus:** Trial transcript of *People vs. Colwell* (p. 28), "Colwell" folder, carton 70, folder 1–3, Edward Oscar Heinrich Papers.

209 **Details about gun:** Block, *The Wizard of Berkeley,* 169.

210 **"I'll get you all":** Block, *The Wizard of Berkeley,* 170.

211 **paraffin wax:** Trial transcript of *People vs. Colwell* (p. 16), "Colwell" folder, carton 70, folder 1–3, Edward Oscar Heinrich Papers.

211 **"These ridges":** Block, *The Wizard of Berkeley,* 165.

211 **"The riflings are made by":** Ibid., 173.

212 **"Caught image on ground glass in focus circa 140mm":** Notes on experiments, "Colwell" folder, carton 70, folder 1–3, Edward Oscar Heinrich Papers.

212 **"There were four panels on this bullet":** Trial transcript of *People vs. Colwell* (pp. 17–19), "Colwell" folder, carton 70, folder 1–3, Edward Oscar Heinrich Papers.

213 **"They match up line for line":** Ibid., 20.

214 **"Will you compute three-eighths of an inch?":** Ibid., 40.

215 **"Absolutely a physical impossibility":** Ibid., 56–57.

215 **"I just wanted to know how":** Ibid., 60.

216 **"I had four hours'":** Ibid., 88.

216 **"His answer was that I couldn't":** Letter from Heinrich to Crossman, March 29, 1926, box 26, folder 6–10, Edward Oscar Heinrich Papers.

216 **"I noticed some of the jurymen":** Ibid.

216 **"Now, if Your Honor please":** Trial transcript of *People vs. Colwell* (pp. 105–7) "Colwell" folder, carton 70, folder 1–3, Edward Oscar Heinrich Papers.

217 **"I found that, in working with this microscope":** Ibid., 105.

217 **"I'd like to see him actually shoot that picture":** Block, *The Wizard of Berkeley,* 177–79.

217 **A commercial photographer was summoned:** Trial transcript of *People vs. Colwell,* "Colwell" folder, carton 70, folder 1–3, Edward Oscar Heinrich Papers.

217 **"The committee agrees that":** National Research Council, *Strengthening Forensic Science in the United States,* 154–55.

217 **Firearms pioneer Calvin Goddard:** "Calvin H. Goddard," *Criminal Justice Law International,* https://criminaljusticelawintl.blog/criminal-justice-law-top-tens-2/top-10-contributors-modern-criminal-justice/calvin-h-goddard.

217 **"I have myself added a comparison":** Letter from Heinrich to Crossman, April 30, 1926, box 26, folder 6–10, Edward Oscar Heinrich Papers.

219 **"Happy birthday, old top":** Letter from Heinrich to Kaiser, December 26, 1926, box 1, John Boynton Kaiser Papers.

219 **"I get my greatest joy in life out of my family":** Letter from Heinrich to Kaiser, May 16, 1927, box 1, John Boynton Kaiser Papers.

219 **"I have been too busy":** Letter from Heinrich to Crossman, November 5, 1929, box 26, folder 6–10, Edward Oscar Heinrich Papers.

220 **"I am returning several book lists":** Letter from Heinrich to Kaiser, September 14, 1928, box 1, John Boynton Kaiser Papers.

220 **"Each had to be shoved through":** Letter from Heinrich to Crossman, August 28, 1929, box 26, folder 6–10, Edward Oscar Heinrich Papers.

220 **"Perhaps I was a little sentimental about it":** Letter from Heinrich to Crossman, August 28, 1929, box 26, folder 6–10, Edward Oscar Heinrich Papers.

220 **"I shall be with you in spirit":** Letter from Heinrich to Theodore, January 14, 1932, 89–44, box 23, file 179, Theodore Heinrich Collection.

221 **"If you don't pay attention":** Letter from Heinrich to Theodore, July 25, 1932, 89–44, box 23, file 179, Theodore Heinrich Collection.

221 **"I cannot avoid feeling somewhat uneasy":** Letter from Heinrich to Theodore, July 5, 1932, 89–44, box 23, file 179, Theodore Heinrich Collection.

221 **"Your grandfather was creative":** Letter from Heinrich to Theodore, July 18, 1932, 89–44, box 23, file 179, Theodore Heinrich Collection.

222 **"I must overcome the fear complex":** Letter from Heinrich to Vollmer, December 23, 1929, box 15, Edward Oscar Heinrich Papers.

222 **"I am persistently and continually beset":** Letter from Heinrich to Bennett F. Davenport, May 18, 1924, box 26, folder 12, Edward Oscar Heinrich Papers.

222 **In 1929, a disastrous string of events:** Smithsonian National Museum of American History; "Social and Cultural Effects of the Depression," Independence Hall Association.

223 **"It would cost investors $30 billion":** "The Economics of World War I," *NBER Digest* (January 2005).

223 **Wickersham Report:** "Wickersham Report on Police," *American Journal of Police Science* 2, no. 4 (July–August 1931): 337–48.

223 **Goddard was soon offered funding:** Details about Northwestern and competition from letter from ibid.

224 **"egotist of the first order":** Letter from Heinrich to Crossman, June 12, 1929, box 26, folder 6–10, Edward Oscar Heinrich Papers.

224 **"The chapters have a surplus of fact":** Review of *Crime's Nemesis,* 84–44, box 10, file 91, Theodore Heinrich Collection.

224 **"Their tendency to over-emphasize":** Letter from Heinrich to Crossman, September 20, 1926, box 26, folder 6–10, Edward Oscar Heinrich Papers.

225 **"I do not entirely agree with your assertion":** Ibid.

225 **He spent months recovering from acute colitis:** Letter from Heinrich to Theodore, September 6, 1932, 89–44, box 23, file 179, Theodore Heinrich Collection.

225 **and saddened from missing his twentieth wedding anniversary:** Letter from Heinrich to Kaiser, September 14, 1928, box 1, John Boynton Kaiser Papers.

CHAPTER 11

227 ***A note about Allene Lamson's cause of death*****:** I've concluded that she died by hitting her head on the sink; I reached this conclusion after reviewing all the pathology reports, appellant statements, and trial transcripts. I examined everything in Heinrich's case files, including his field notes and details about his experiments. I consulted with several experts in the field. That being said—it still might have been murder, as you'll see at the end of Chapter 11.

227 **She squirmed in her chair:** "Lamson Case Is Continued for Ten Days," *Healdsburg Tribune* (CA), June 5, 1933; as well as several press photos featuring the little girl from the arraignment hearing; scenes also from various photos from the International News Photos agency on the day of the arraignment.

229 **"I've just been figuring expenses":** Letter from Theodore to Heinrich, February 2, 1932, box 27, folder 25–26, Edward Oscar Heinrich Papers.

229 **"I fear you are still":** Letter from Heinrich to Theodore, January 16, 1933, 89–44, box 23, file 179, Theodore Heinrich Collection.

229 **"Say to Mort that":** Letter from Heinrich to Marion, May 12, 1932, 89–44, box 23, file 179, Theodore Heinrich Collection.

229 **Inside his home laboratory:** "Clippings, articles, publications" in Lamson Murder Case Collection.

230 **a how-to manual for calculating:** *Simplified Blood Chemistry as Practiced with the Ettman Blood Chemistry Set* comes from a catalogue of Heinrich's books at UC Berkeley.

230 **joint forensic tests:** *Supreme Court of the State of California,* 188–96.

230 **Proescher's background:** Stephen J. Morewitz and Mark L. Goldstein, *Handbook of Forensic Sociology and Psychology* (New York: Springer, 2014); Frederick Proescher, "A Remarkable Case of Carcinoma of the Gall Bladder," *JAMA* 48, no. 6 (1907): 481–83; 1940 California census confirms 1878 birth in Germany, https://www.ancestry.com/1940-census/usa/California/Frederick-Proescher_2ghgfj.

231 **Stains inside nurse's closet and David's work shirt:** *Supreme Court of the State of California,* 188–90.

231 **they were spots of varnish:** ibid., 204–5.

231 **Items in the bonfire:** ibid., 204–23.

231 **"I made more than one test":** ibid., 188, 196–97.

231 **"Dr. Proescher took the sample and cut off about":** ibid., 197–98.

231 **"Leucomalachite Green test":** "Leucomalachite Green Presumptive Test for Blood," National Forensic Science Technology Center, https://static.training.nij.gov/lab-manual/Linked%20Documents/Protocols/pdi_lab_pro_2.18.pdf.

231 **Oscar's theory of body positioning:** *Supreme Court of the State of California,* 33–34.

232 **spots of arterial blood:** Russell and Winters, *The Case of David Lamson,* 34.

232 **"Measurements in a direct line from the point":** Russell and Winters, *The Case of David Lamson,* 34.

233 **No blood cast-off and other results from experiments:** "Experiment Summary," dated June 20, 1933, in the "Lamson" folder, carton 71, folder 31–41, Edward Oscar Heinrich Papers.

233 **Definition of hemolyzed blood:** Mark Okuda and Frank H. Stephenson, *A Hands-On Introduction to Forensic Science: Cracking the Case* (New York: Routledge, 2019), 94.

235 **Even newspaper writers:** "Battle of Scientists Centers Around Dave," *Oakland Post-Enquirer,* September 5, 1933; "Heinrich Applies Benzedine Test on Floor," *San Francisco Call Bulletin,* June 24, 1933.

236 **"Heinrich, a noted criminologist":** Associated Press of San Francisco, August 28, 1933.

236 **"nationally known criminologist and ace witness":** "Blood Tests Disputed in Lamson Trial," *Arizona Republic,* September 9, 1933.

236 **"For years I have been enchained":** Letter from Heinrich to Kaiser, October 9, 1922, box 28, folder 14, Edward Oscar Heinrich Papers.

237 **DA swings the pipe:** *Supreme Court of the State of California,* 306.

237 **The affair with the nursemaid:** Russell and Winters, *The Case of David Lamson,* 16–17, 91–92; "Husband to Face Murder Charge in Campus Mystery," *Fresno Bee* (CA), June 1, 1933.

237 **"I know nothing of her private affairs":** "Girl to Take Stand in Effort to Save Life of Stanford Man," *Oakland Tribune,* August 21, 1933.

238 **Sara Kelley and David collaborated:** Russell and Winters, *The Case of David Lamson,* 17–22; *Supreme Court of the State of California,* 248–52.

238 **"Everyone knows that if the husband starts to go":** Ibid., 252.

238 **There were other theories from the district attorney:** "Discord in Lamson Home," *Oakland Tribune,* August 31, 1933.

239 **David demanded sex:** Russell and Winters, *The Case of David Lamson,* 82; *Supreme Court of the State of California,* 167.

239 **Almost twenty years earlier:** Russell and Winters, *The Case of David Lamson,* 14; "Accused Man's Mother Tells Story of Boyhood Fun Tragedy in Which Son Was Absolved," *Santa Cruz Evening News* (CA), June 5, 1933.

239 **"Mother's Day with silk hose":** "No Hint of Trouble in Home Told by Journal," *Oakland Tribune,* June 11, 1933.

240 **"Suddenly the betrothal was broken":** "Romance of Mrs. Lamson Well Known," *Oakland Tribune,* June 3, 1933.

240 **"The benzidine test was negative":** *Supreme Court of the State of California,* 224–26.

240 **He noticed a disturbing pattern:** Ibid., 245–46.

241 **"Heinrich is meticulous of speech":** "Lamson Jury Casts One Ballot—to See Movie," *Oakland Tribune,* September 1, 1933.

241 **"I understand English":** Ibid.

241 **"Do you doubt my integrity?":** Ibid.

242 **"One knock would crush the skull":** *Supreme Court of the State of California,* 413.

242 **"What was your wife's position in the tub":** "Lamson Tells Own Story," *Oakland Tribune,* September 7, 1933.

243 **"The injuries are typical of what we call":** "Blood Tests Disputed in Lamson Trial," *Arizona Republic,* September 9, 1933.

243 **"Do you say it is impossible that those wounds":** *Supreme Court of the State of California,* 302.

243 **"When blood is charred completely":** "Blood Tests Disputed in Lamson Trial."

244 **"Doctor, did you perform an experiment at the Lamson cottage":** *Supreme Court of the State of California,* 266–67.

244 **"The only way to prove it would be to kill someone":** "Judge Halts Attempt of Dr. Heinrich," *Madera Daily Tribune* (CA), September 12, 1933.

244 **Heinrich can't testify to blood spatter:** *Supreme Court of the State of California,* 286.

245 **Oscar had sliced an artery:** "Ask Mr. Heinrich," *True Detective,* August 1944, 111.

245 **"We are not going to let this man walk off to go out of here":** *Supreme Court of the State of California,* 313.

245 **"Why, you couldn't fracture":** Ibid., 307.

245 **Nelle Clemence was anxious:** "Sketches of Jurors Given," *Oakland Tribune,* September 16, 1933; "Fate of David A. Lamson Rests in Hands of These Men and Women," *Oakland Tribune,* August 25, 1933.

245 **Details about David Lamson's jury deliberations:** Russell and Winters, *The Case of David Lamson,* 91–92; "Lamson Convicted Himself—Juror," *Healdsburg Tribune* (CA), September 18, 1933; "Sentence to Death Tuesday," *Nevada State Journal,* September 18, 1933; "2 Lamson Jurors Served on Jury in Matlock Case," *Oakland Tribune,* September 18, 1933.

246 **"One can imagine":** "Five Women, Seven Men to Try Case," *Oakland Tribune,* August 25, 1933.

246 **"My mind was made up almost a week before":** "Lamson Convicted Himself—Juror."

247 **Juror tampering allegations:** "Attorneys to Impeach Lamson Jury," *Oakland Tribune,* September 23, 1933.

247 **flipped through newspapers during a picnic:** "Attorneys to Impeach Lamson Jury."

247 **"You know what happened":** Ibid.

247 **Joe Matlock had been convicted of gunning down:** "The Holdout in the Matlock Case," *Mercury News* (San Jose, CA), May 26, 2012.

247 **After the meal:** *Oakland Tribune,* September 23, 1933.

247 **George Peterson's misconduct:** *Supreme Court of the State of California,* 589–607.

247 **Jury experiments:** Ibid., 585–89; Russell and Winters, *The Case of David Lamson,* 91–93.

249 **"There was evidence of attempts to wash out the blood":** "Trial Errors, Jury's Misconduct Charged in Lamson's Plea," *San Jose Evening News* (CA), September 23, 1933.

250 **"I know David Lamson is guilty":** *Supreme Court of the State of California,* 571–81.

250 **"We are in for a critical winter":** Letter from Heinrich to Theodore, November 4, 1933, 89–44, box 23, file 180, Theodore Heinrich Collection.

251 **"The steady drain":** Letter from Heinrich to Theodore, November 1, 1934, 89–44, box 23, file 180, Theodore Heinrich Collection.

251 **"A good wife during such a period":** Letter from Heinrich to Theodore, December 10, 1934, 89–44, box 23, file 180, Theodore Heinrich Collection.

251 **"Please do not allow":** Letter from Heinrich to Theodore, May 21, 1935, 89–44, box 23, file 180, Theodore Heinrich Collection.

252 **The scene was appalling:** "Lawyers Swap Punches in Lamson Hearing Row," *Los Angeles Times,* September 26, 1933; "Lamson Hearing Marked by Genuine Fist Fight Between Two Attorneys," *Santa Cruz Sentinel* (CA), September 25, 1933.

252 **Details about affidavits:** "Trial Errors, Jury's Misconduct Charged in Lamson's Plea"; *Supreme Court of the State of California,* 431–608.

252 **"I was always convinced of Lamson's innocence":** "Irregularities in Lamson Jury Said Cause for Retrial," *Santa Cruz Evening News* (CA), September 23, 1933.

252 **"Are you calling me a perjurer?":** "Lawyers Stage Fist Fight at Lamson Hearing," *Mercury Herald* (San Jose, CA), September 26, 1933.

252 **"the most shocking, the most hideous verdict":** "Trial Errors, Jury's Misconduct Charged in Lamson's Plea."

253 **"Don't worry, everything":** "Lamson Hearing Marked by Genuine Fist Fight Between Two Attorneys."

253 **Details about hearing for death sentence:** "'I'm Innocent,' Lamson Asserts as Judge Sets Execution Date," *Mercury Herald* (San Jose, CA), September 27, 1933.

253 **"My boy doesn't wear his heart on his sleeve":** "Dave Didn't Kill Allene, Says Mother," *Oakland Tribune,* October 1, 1933.

254 **had petitioned to adopt the little girl:** "Conviction of Lamson Fails Dash Composure Open Fight for Child," *Madera Daily Tribune* (CA), September 18, 1933.

254 **"Allene [when she was alive] asked me to take care of little":** "Lamson's Mother Tells of His Life with Allene as Proof of Innocence," *Oakland Tribune,* October 1, 1933.

254 **"The verdict means nothing to me":** "Justice Miscarried, Heinrich Declares," *Oakland Tribune,* September 18, 1933.

254 **"Any intelligent person":** "Lamson Held Guiltless by Vollmer," *Oakland Tribune,* February 12, 1934.

255 **Details about Lamson's trip to San Quentin:** "Lamson Placed in 'Death Row,'" *San Bernardino Daily Sun,* October 7, 1933; "Prison Concessions Granted to Lamson," *San Bernardino Daily Sun,* October 8, 1933.

255 **Custody battle:** "Custody of Lamson Baby Fight Delayed," *Madera Daily Tribune* (CA), October 24, 1933.

256 **"It exemplifies the frightful potentialities":** Russell and Winters, *The Case of David Lamson,* 16.

256 **Supreme Court decision:** *Supreme Court of California statement PEOPLE v. LAMSON.*

257 **"We brought into our chambers the wash basin":** "Lamson Foresees Freedom Within Month, Daughter Still Unaware that Mother Is Dead," *Oakland Tribune,* October 14, 1934.

258 **"She still believes that her mother has":** Ibid.

258 **"We have never changed our opinion":** "Highlights in Lamson Case from Tragedy Until Today," *Oakland Tribune,* October 14, 1934.

258 **denied the request and ordered a new trial:** "Lamson to Get New Trial on Charge of Murdering His Wife," *Healdsburg Tribune* (CA), December 7, 1934.

258 **"I'm so happy":** "David Lamson Back in Santa Clara Co. Jail," *Santa Cruz Sentinel* (CA), November 15, 1934.

258 **"When I passed this way before":** "Lamson to Learn Fate," *Oakland Tribune,* November 15, 1934.

259 **Homecoming dinner:** Ibid.

259 **thirty thousand people a year died:** "Ordeal Is Ended for David Lamson," *Oakland Tribune,* May 15, 1950.

259 **"McKenzie immediately—and by no means silently":** Ibid.

259 **replica bathroom:** "Bathtub, Girl Assist Lamson," *Petaluma Argus-Courier* (CA), April 12, 1935.

259 **The jury deadlocked 9–3:** "Jury Deadlocked," *Daily News* (NY), March 24, 1936.

260 **Heinrich's experiments:** "Expert Heard: Lamson Fall Tests Cited," *Los Angeles Times,* April 3, 1935.

260 **"They say that when a high-velocity bullet":** "Accused Tells of Reactions to Jury's Death Decree in Wife's Death," *Pittsburgh Press,* July 8, 1935.

260 **his book was published:** "Lamson Writes of Men in Death Row Where He Lived," *Missoulian* (MT), September 20, 1935.

260 **"The frog at the bottom of the well thinks":** Lamson, *We Who Are About to Die,* x–xi.

260 **"It is really an appeal to consciousness not to mercy":** "A Compelling Revelation of Life in the Death House," *New York Times,* October 6, 1935.

260 **Alexander Woollcott, a legendary book critic:** Display ad 63, *New York Times,* January 12, 1936.

261 **November mistrial:** "Court Grants Lamson Plea for Mistrial," *Daily News* (NY), November 24, 1935.

261 **Syer disqualifies himself:** "Disqualifies Self as Judge for David Lamson's 3rd Trial," *Daily Inter Lake* (Kalispell, MT), November 2, 1935.

261 **fourth trial:** "State Debates Trying Lamson Fourth Time," *Daily News* (NY), March 25, 1936.

261 **"With tireless effort we sought to gather legal evidence":** "Former Stanford Campus Leader Is Turned Loose After Fourth Trial for Killing Wife Is Deadlocked," *Dayton Herald,* April 3, 1936.

262 **Lamson reunited with Bebe:** "David Lamson Freed from Jail as Thrice-Tried Murder Case Is Dismissed by Prosecutor," *Oakland Tribune,* April 3, 1936.

262 **Details about his life after prison:** "David Lamson's Ordeal," *San Francisco Examiner,* May 28, 2017; "David Lamson, Tried 4 Times for Murder," *New York Times,* August 9, 1975; "The 1933 Lamson Case at Stanford: A Murder?," *Mercury News* (San Jose, CA), March 12, 2017.

262 **"He never discussed the specifics of the case":** Bernard Butcher, "Was It Murder?," *Stanford Alumni Magazine,* January/February 2000.

263 **"I looked up at the Row. I waved back to them":** Lamson, *We Who Are About to Die,* 268.

263 **"As you no doubt have observed in the Lamson case":** Letter from Heinrich to Kirk, June 10, 1935, box 12, folder 49, Edward Oscar Heinrich Papers.

263 **Paul Kirk background:** Douglas O. Linder, "Selected Testimony of Doctor Paul Kirk in Sam Sheppard's 1966 Murder Trial," *Famous Trials,* https://www.famous-trials.com/sam-sheppard/12-excerpts-from-the-trial-transcripts/24-kirktestimony; "Suspects May Be Released Here," *Arizona Daily Star,* November 10, 1935.

264 **Sam Sheppard case:** Linder, "Selected Testimony of Doctor Paul Kirk"; "Kirk Investigation Photos," "The Sam Sheppard Case: 1954–2000," Cleveland State University online database, https://library.csuohio.edu/ehs/access-database.

265 **"One can tell, for example":** National Research Council, *Strengthening Forensic Science in the United States,* 178.

EPILOGUE

267 **August Vollmer, Oscar's colleague at UC Berkeley:** Dinkelspiel, "Remembering August Vollmer, the Berkeley Police Chief Who Created Modern Policing."

267 **Calvin Goddard was responsible for a number of:** "Goddard, Calvin Hooker," *World of Forensic Science,* Encyclopedia.com (Access Date), https://www.encyclopedia.com/science/encyclopedias-almanacs-transcripts-and-maps/goddard-calvin-hooker.

267 **Dr. Paul Kirk helped Vollmer establish:** "Kirk, Paul Leland," *World of Forensic Sciences,* Encyclopedia.com, https://www.encyclopedia.com/science/encyclopedias-almanacs-transcripts-and-maps/kirk-paul-leland.

267 **Details about recommendations:** National Research Council, *Strengthening Forensic Science in the United States,* 6.

268 **In 1932, the FBI's first director:** "The FBI Laboratory: 75 Years of Forensic Science Service," Federal Bureau of Investigation, *Forensic Science Communications* 9, no. 4 (October 2007).

269 **"The investigation found":** "Report Criticizes Scientific Testing at F.B.I. Crime Lab," *New York Times,* April 16, 1997.

269 **wrongful convictions:** "The Causes of Wrongful Conviction," The Innocence Project, https://www.innocenceproject.org/causes-wrongful-conviction.

269 **Bloodstain pattern:** Pamela Colloff, "Blood Will Tell," *New York Times Magazine,* May 31, 2018.

270 **The academy suggested that the advent:** "Science Found Wanting in Nation's Crime Labs," *New York Times,* February 4, 2009.

270 **independent federal agency:** Ibid., 19.

270 **Daubert Hearings:** "What Is a Daubert Hearing?" Office of Medical and Scientific Justice.

270 **Cameron Todd Willingham:** "Cameron Todd Willingham: Wrongfully Convicted and Executed in Texas," The Innocence Project, September 13, 2010.

270 **In 2014, a former crime scene analyst:** Radley Balko, "A Crime Lab Analyst Killed Himself after Contamination Wrongly Made Him a Suspect in a 30-Year-old Murder," *Washington Post,* June 5, 2017.

271 **false confessions:** "False Confessions & Recording of Custodial Interrogations," The Innocence Project, https://www.innocenceproject.org/false-confessions-recording-interrogations.

272 **"Don't know whether I've been tossed":** Letter from Heinrich to Kaiser, November 11, 1950, box 1, in John Boynton Kaiser Papers.

272 **Mort's military history:** Letter from Heinrich to John McCloy, January 23, 1946, carton 85, folder 159, Edward Oscar Heinrich Papers.

272 **Theo's Bronze Star and career:** "Biographical Sketch" (p. 6), 89–44, box 23, file 180, Theodore Heinrich Collection.

272 **Monuments Men:** *The Monuments Men Foundation*, https://www.monumentsmenfoundation.org/the-heroes/the-monuments-men.

272 **Both of his sons followed Oscar's own career:** Mort's background from various family letters and obituary for "Mary Elizabeth 'Betty' Onthank Heinrich," *Honolulu Advertiser,* October 5, 2001.

273 **"I am still waiting for you":** Letter from Heinrich to Theodore, August 12, 1953, 89–44, box 24, file 188, Theodore Heinrich Collection.

273 **Stroke in 1952:** Letter from Marion to Heinrich, October 13, 1952, box 2, John Boynton Kaiser Papers.

273 **Hypertension:** Letter from Heinrich to Theodore, May 14, 1953, 89–44, box 24, file 188, Theodore Heinrich Collection.

273 **fifty-five hours a week:** Letter from Heinrich to Theodore, February 15, 1949, 89–44, box 24, file 187, Theodore Heinrich Collection.

273 **"I am annoyed by nothing":** Letter from Heinrich to Theodore, August 27, 1953, 89–44, box 24, file 188, Theodore Heinrich Collection.

273 **"You will find as you grow in your field":** Letter from Heinrich to Mortimer, March 26, 1947, box 11, folder 19, Edward Oscar Heinrich Papers.

273 **"I regard my greatest accomplishment":** Letter from Heinrich to Theodore, June 6, 1948, 89–44, box 24, file 186, Theodore Heinrich Collection.

274 **"Best wishes. Get well soon":** Telegram from Kaiser to Heinrich, September 29, 1953, box 1, John Boynton Kaiser Papers.

274 **Death:** "Heinrich Rites Today," *San Francisco Examiner,* September 30, 1953.

274 **"America's greatest forensic scientist":** Evans, *Murder 2,* 112–13.

INDEX

Adam, Elizabeth, 164–65, 167, 182
Alameda County, California, 37
Arbuckle, Roscoe "Fatty"
  background of, 88
  and film tribute to Heinrich, 128
  hung juries' acquittals of, 117, 120, 125
  and ice allegation, 107, 112, 116
  in jail, 94–95
  murder charges against, 93–94
  and police investigation of Rappe's death, 90–91, 92–93
  post-trial life of, 127–28
  and Prevon's testimony, 91–93, 104–5, 106–8
  ruined career of, 102–3, 106, 126, 128
  second and third retrials of, 122–23, 125, 126
  trial testimony of, 115–16
  See also Arbuckle murder case
Arbuckle murder case
  and cause of Rappe's death, 103, 106
  and fingerprint evidence, 98, 113–15, 117, 122
  and grand jury trial, 102–9
  and hair analysis, 99, 113
  Heinrich's distress following outcome of, 159, 219
  Heinrich's investigation of, 95–98, 101
  Heinrich's trial testimony for, 101, 112–14, 116–17, 122–23
  and housekeepers' testimony, 101, 108–9, 112–13
  and hung juries' acquittals of Arbuckle, 117, 120, 125
  and ice allegation, 107, 112, 116
  and inconsistency in statements of witnesses, 102
  and jail time of Arbuckle, 94–95
  media coverage of, 101, 106
  and murder charges against Arbuckle, 93–94
  police investigation of, 90–91, 92–93
  and Prevon's testimony, 91–93, 104–5, 106–8
  and problematic witnesses, 98–99, 104–5, 107–8, 112
  and second and third retrials of Arbuckle, 122–23, 125, 126
  and trial of Arbuckle, 111–14, 115–17
  and Vollmer's testimony, 123

ballistics
  and Colwell murder case, 209, 212–18, 268
  comparison microscopes used in, 210–11, 212, 215–16, 218, 268
  contemporary standards for use of, in evidence in trials, 218
  Heinrich's innovations in, 218
  origins of field, 210
  photographic evidence used in, 212, 215–17, 268
Barbe, Gilbert Warren, 181, 183
Barnett, Frank, 197, 200–201, 204
Bates, Sid, 139, 142–43
Bender, David "Kid," 94–95
Berkeley Police Department, 73–74, 77, 188
Black Kit Bag, The (Heinrich), 152, 154, 159–60

Index

Blake, Alice
  and district attorney's case against Arbuckle, 98
  and investigation of Rappe's death, 90, 91, 93
  placed in protective custody, 105
  tending to Rappe, 106
  trial testimony of, 108, 111
blood pattern analysis (BPA)
  contemporary perspectives on, 265, 268
  and Ferguson's dismembered ear case, 187, 203
  Heinrich's expertise in, 20, 208, 264
  and Lamson investigation, 20–21, 22, 230–33, 244–45, 249–50
  and mismanaged crime scenes, 21
  NAS report on, 265
  and Schwartz murder investigation, 176, 180
  taught by Heinrich at Berkeley, 201
  use of ultraviolet light in, 208
Boulder, Colorado, 38
Bow Street Runners, 210
Boyle, Salome, 95–96, 98, 122
Brady, Matthew
  and acquittal of Arbuckle, 117
  and case against Arbuckle, 102, 103, 122
  and Heinrich, 95, 112
  key witnesses of, 98, 102, 108, 109, 122
  and trial/retrials of Arbuckle, 104–9, 112, 122
Brennan, Kate, 112–13
Bridges, Herbert, 243
Brown, Kevin, 270–71
Brown, Viola, 248
Bureau of Forensic Ballistics in New York City, 210
Bureau of Investigation, 4

California State Supreme Court, 254, 256–58
Capone, Al, 47, 223
Case of David Lamson report, 256
Christmas holidays, 119–22
circumstantial evidence, 4, 114
Clemence, Nelle, 245–50, 252
clothing analysis
  and Ferguson's dismembered ear case, 189
  and Siskiyou train robbery case, 147–51
collecting/tracking tendencies of Heinrich
  evidence collected, 155
  and father's death, 154–55
  field journals, 96
  financial logs, 41, 154–56
  household matters, 156
  logs of lab assistants, 96, 155
  newspaper clippings and periodicals, 155
  and obsessive-compulsive nature of Heinrich, 1–3
  stamp collecting, 40
  tracking urine levels, 156
  and UC Berkeley's archive from Heinrich's lab, 1–3
Colma, California, 43–44
Colwell, Martin
  arrest of, 209
  conviction and sentence of, 218
  and death of McCarthy, 207–8
  defense team of, 208, 213–14
  trials of, 212–18
Colwell murder case
  arrest of Colwell, 209
  ballistics in, 209, 211–18, 268
  death of McCarthy, 207–8
  Heinrich's investigation of, 208, 209, 211–12
  Heinrich's trial testimony in, 212–14, 216–17
  and McGovern's testimony, 214–15
  and photo evidence for juries, 212–14, 218, 243, 268
  and trials of Colwell, 212–18
comparison microscopes used in ballistics, 210–11, 212, 215–16, 218, 268
confessions, false, 271
Coolidge, Calvin, 133
corruption in 1920s, 4
Crandall, Harry, 102
crime lab of FBI, 269
crime rates, 47, 223
Crime's Nemesis (May), 224

Index

Cumberland, Vera, 103
Curse of the Gleaming Eye, The (Heinrich), 153

Daubert standard, 77, 270
Daugherty, Harry, 99
DeAutremont, Hugh
- and attempted train robbery, 130–31, 138–43
- background of, 134
- capture and imprisonment of, 160, 161
- and commitment to heist, 133–34
- planning and preparing for heist, 135–38
- trip to see father, 137
- *See also* Siskiyou train robbery case

DeAutremont, Paul, 134, 137, 158
DeAutremont, Ray
- and attempted train robbery, 131, 138–43
- background of, 134–35
- capture and imprisonment of, 160–61
- planning and preparing for heist, 135–38
- previous prison term of, 135
- *See also* Siskiyou train robbery case

DeAutremont, Roy
- and attempted train robbery, 129–31, 138–43
- background of, 134–35
- capture and imprisonment of, 160–61
- girlfriend of, 137, 138
- and Heinrich's description of suspect, 152
- and Heinrich's investigation of heist, 157–58
- knee injury of, 137–38, 139
- mail receipt found in overalls of, 157–58
- murders committed by, 140, 142
- planning and preparing for heist, 135–38
- *See also* Siskiyou train robbery case

decomposition of bodies, 188
Delmont, Maude
- and Arbuckle's testimony, 116
- and coroner's inquest, 103–4
- and district attorney's case against Arbuckle, 98
- and grand jury trial, 109
- and ice claim, 107
- inconsistency in statements of, 102
- and police investigation of Rappe's death, 92, 93
- and Prevon's testimony, 107–8
- tending to Rappe, 87, 88, 106
- and testimony of other witnesses, 105

Dempsey, Jack, 40
detective stories authored by Heinrich, 152–54, 159–60, 193
DNA testing, fallibility of, 271
domestic violence, 118
Dominguez, Frank, 107–8, 111
Dougherty, Elvyn, 140, 143
Doyle, Arthur Conan, 7
Dunbar, Louise, 11
Dunnell, Leo, 216
Durfee, Minta, 89, 127–28

Edson, Philips, 76
Eisenschimmel, Carl, 53–55, 57, 72, 229
Emig, William, 18
Emmons, E. J., 78, 79
entertainment industry and Hollywood, 90–91, 99–100, 102, 118, 126. *See also* Arbuckle murder case
ethnicity, conclusions about, 152
eyewitnesses, misidentification by, 75–76, 271

Federal Bureau of Investigation (FBI) crime lab, 269
Ferguson, Bessie
- background of, 195–97
- discovery of ear, 185–86
- lifestyle of, 195–97, 200
- recovery and identification of body, 193–95
- *See also* Ferguson murder case

Ferguson murder case
- blood analysis in, 187, 203
- cover up of, 198
- discovery of ear, 185–86
- forensic entomology employed in, 188–89, 268

Ferguson murder case *(cont.)*
  Heinrich's investigation of, 186–92, 194–95, 197–98, 199–201, 203–5
  identification of body, 194–95
  media coverage of, 189–90, 192
  police investigation of, 192
  and profiling killer, 197–98
  recovery of body, 193–94
  suspect list in, 200–201, 204–5
  unsolved status of, 205
finances of Heinrich
  and Christmas holidays, 120–21
  and detailed financial logs, 41, 154–56
  and father's financial failings, 41, 42, 120, 221, 222
  and financial support for mother, 124, 155–56, 169–70
  income decline during Depression, 225, 228–29
  and mortgage foreclosure, 41–42
  and opening of new office, 71
  and periods of underemployment, 225
  stabilization of, 209
  struggles of Heinrich with, 156–57, 169–70, 220, 221–22, 273
  and Theodore's finances in later life, 272–73
  and Theodore's studies abroad, 220, 221–22, 229, 251
  wife's ignorance of, 121, 184
fingerprints and fingerprinting science
  and Arbuckle murder investigation, 98, 113–15, 117, 122
  contemporary perspectives on, 115
  history of, 114
  and poroscopy, 114
  and Schwartz murder investigation, 172, 178
fire investigation, 174–76. *See also* Schwartz murder case
Fischer, John H., 210
Fitzgerald, John P., 238, 244
forensic science
  calls for standardization in, 269, 270
  Daubert standard for, 77, 270
  development of disciplines within, 208
  early perceptions of, 23
  and emergence of independent forensic labs, 223–24
  forensic entomology, 188–89, 268
  forensic geology, 67–70, 190–92, 267–68
  growing pains in field of, 208
  innovations in, 267
  juries' reactions to, 117, 144–45, 212, 217, 219, 263, 271
  NAS recommendations for field, 269, 270
  testing and reviewing techniques in, 270
  training and qualifications of experts in, 269
  and wrongful convictions, 269, 270, 271
  See also specific disciplines, including fingerprints and fingerprinting science
forensic science career of Heinrich
  and ballistics, 33–34, 209–17, 218
  and burden of criminal investigations, 121–22
  credibility as expert witness, 127
  and detective stories authored by Heinrich, 152–54, 159–60
  dissatisfaction of Heinrich with, 121–22, 159, 184
  and effects of Great Depression, 228
  expansion of firm's services, 228
  as first crime scene investigator, 5
  and forensic entomology, 188–89, 268
  and forensic geology, 67–70, 190–92, 267–68
  general cases, 170
  and hair-related evidence (*see* hair analysis)
  and handwriting (*see* handwriting analysis)
  and hubris of Heinrich, 224–25
  impact of, 1, 5–6, 267–68, 274
  innovation of Heinrich in, 208, 212, 218
  as instructor at UC Berkeley, 36–37, 38, 201, 268
  juries' responses to evidence presented by, 117, 125, 144–45, 212, 217, 219, 234, 263, 271

media coverage of, 21–22, 170, 184, 190, 198, 224, 236
and microscopes of Heinrich, 3, 210–11, 216–17, 218–19, 268
and missteps in trials, 22–23
and notoriety of Heinrich, 159, 160, 161, 184
and petrographic analysis, 67–70, 80–81, 189, 190–92
and photo evidence for juries, 212–19, 234, 243–44, 268
and profiling, 40, 59–60, 197–98
and publications in the field, 171, 273
and relationships with other forensic specialists, 202–3
and reputation of Heinrich, 5, 21, 22, 208–9, 218–19, 274
and rivals, 38, 51, 55, 155, 223–24, 273 (*see also* McGovern, Chauncey)
and Sherlock-Holmes comparisons, 21–22, 159
and stiff presentation style, 22–23, 236–37, 241
time demands of, 219–20, 273
and trace evidence analysis, 70, 149, 268
and traveling for work, 156, 170, 219, 229
See also Arbuckle murder case; Colwell murder case; Ferguson murder case; Heslin murder case; Lamson murder case; Schwartz murder case; Siskiyou train robbery case
Frye v. United States, 77

Galton, Sir Francis, 114
geology, forensic, 67–70, 190–92, 267–68
Ghetti, Leonora, 250
Goddard, Calvin
expertise in ballistics, 210, 267
and Heinrich's use of photographic evidence in court, 218
and Northwestern lab, 224
and St. Valentine's Day Massacre in Chicago, 223
Gonzalez, Walter, 167–68
Graham, Iva, 187
graphology, 52–55
Gravelle, Philip, 210
Great Depression
and Heinrich's business, 228
impact of, 8–9, 223
stock market correction preceding, 222, 223
years leading up to, 169
Great Gatsby, The (Fitzgerald), 168
Great Train Robbery, The (film), 132
Gross, Hans, 69

hair analysis
and Arbuckle murder investigation, 99, 113
and Schwartz murder investigation, 178
and Siskiyou train robbery case, 148, 152, 158
and taxicab driver murder case, 40
handwriting analysis
contemporary perspectives on, 268
and Dempsey's draft-dodging trial, 40
graphology compared to, 52–53
and Heinrich's pharmacy work, 30
Heinrich's pioneering role in, 267
and Heinrich's studies with Kyka, 37
and Heslin murder case, 50–51, 55–57, 68, 81, 84
history of, 51–52
NAS recommendations for use of, 52
and Schwartz murder investigation, 181
and Siskiyou train robbery case, 158
taught by Heinrich at Berkeley, 201
used in trials, 51–52
Hanna, Edward, 47–48, 49, 57–58, 67
Harding, Warren, 46, 133
Harliss, J. A., 248
heart balm lawsuits, 165
Hegerich, George, 248, 249
Heinrich, Adalina Clara (sister), 25
Heinrich, Albertine (mother)
death of, 170
marriage and family of, 25–26
Oscar's financial support of, 124, 155–56, 169–70
Oscar's respect for teachings of, 26
and suicide of husband, 27–28, 169

Index

Heinrich, Anna Matilde (sister), 26
Heinrich, August (father)
  financial failings of, 27, 41, 120, 221
  and financial record keeping of Oscar, 154–55
  marriage and family of, 25–26
  relocation of family to Tacoma, 26
  suicide of, 27–29, 42, 169
Heinrich, Edward Oscar
  anxiety of, 38, 71–72
  birthday reflections of, 126–27
  chemistry training of, 30, 31–32, 173–74
  children of, 127, 220 (*see also* Heinrich, Mortimer; Heinrich, Theodore)
  and Christmas holidays, 120–22
  death of, 1, 274
  doubts and insecurities of, 39, 121, 222, 236
  early employment of, 32–33, 37, 38
  education of, 27, 29, 31–32
  and family life, 156, 171–72, 219, 229
  and father's financial failings, 41, 120, 221, 222
  and father's suicide, 28–29, 169, 221
  health of, 225, 273, 274
  hobbies of, 153
  lab of (*see* Heinrich Technical Laboratories)
  marriage of, 32, 82, 198–99 (*see also* Heinrich, Marion Allen)
  methodical nature of, 96, 98 (*see also* collecting/tracking tendencies of Heinrich)
  money worries of (*see* finances of Heinrich)
  obsessive-compulsive nature of, 2, 29, 38
  pharmaceutical training of, 29–30
  physical appearance of, 3
  writing ambitions of, 152–54, 159–60, 193
  youth of, 25–31, 120
Heinrich, Gustav Theodor (brother), 25–26
Heinrich, Marion Allen (wife)
  children of, 38, 121
  and family vacation, 171–72
  marriage of, 32, 82, 198–99
  and money struggles of Oscar, 121, 184
  and Oscar's travel for work, 156
  police protection for, 229
  and Rappe murder case, 123–24
  sensitive nature of, 121
Heinrich, Mortimer (son)
  career of, 272
  children of, 273
  and donation of Oscar's lab contents to Berkeley, 1–2
  and family vacation, 171–72
  and financial struggles of family, 220
  Heinrich's relationship with, 127, 220, 273–74
  and Oscar's travel for work, 156
  personality of, 193
  and police protection for mother, 229
  at University of Oregon, 251
  WWII service of, 272
  youth of, 38, 121, 193
Heinrich, Theodore (son)
  career of, 272
  and family vacation, 171–72
  and financial strain on father, 220, 221, 229, 251, 272–73
  Heinrich on romances of, 251
  Heinrich's pride in, 192–93, 220–21
  Heinrich's relationship with, 127, 193, 220, 272–74
  personality of, 192–93
  studies and travels abroad, 220, 221–22, 229, 251
  writing success of, 220
  WWII service of, 272
  youth of, 38, 121, 192–93
Heinrich Technical Laboratories
  as earliest general forensics lab, 5, 21
  and financial struggles of Heinrich, 41, 42
  library in, 3–4
  opening of, 33
  organization of, 37–38
  questioned-documents business of, 51, 52
  and second lab considered, 42
  UC Berkeley's archive from, 1–3
Herschel, Sir William James, 114
Heslin, Patrick, 44–46, 47–48, 67

Index

Heslin murder case, 44–45
and coroner's report, 66
and eyewitness testimony, 75, 78–79
graphologists consulted in, 53–55, 57, 72
handwriting analysis in, 50–51, 55–57, 68, 81, 84
Heinrich's investigation of, 48–49, 53–57, 66, 67–70
Heinrich's trial testimony in, 80–81
media coverage of, 50, 65, 75, 80
and petrographic analysis of sand, 67–70, 80–81
police investigation of, 65–67
and polygraph test of Hightower, 72–78
and ransom notes, 48–49, 50–51, 53–57, 68, 84
and recovery of Heslin's body, 58–63
and trial of Hightower, 78–81, 84–86
Hicks, Charlie, 152
Hightower, William A.
conviction and sentence of, 84–86
and corpse of Father Heslin, 58–63
death of, 86
mental illness of, 70–71, 85–86
trial of, 78–81, 84–86
See also Heslin murder case
Hindu Ghadar Conspiracy, 39–40
Hines, William, 60, 65
Hollywood, 90–91, 99–100, 102, 118, 126. *See also* Arbuckle murder case
Holmes, Sherlock, comparisons made to, 21–22, 159
homicide rates, 4, 133
Hoover, Herbert, 8
Hoover, J. Edgar, 268–69

India, anti-British rebellion in, 39–40
Industrial Workers of the World (IWW), 135
Innocence Project, 76, 269
insects in crime solving, 188–89, 268
In the Chapel (Heinrich), 153

Jack the Ripper, 56, 197
Jameson, Jean, 103
Johnson, Charles "Coyle," 141
juries
and challenges faced in jury rooms, 246
difficulties with forensic evidence, 117, 144–45, 212, 217, 219, 263, 271
hung, 117, 120, 125
in Lamson murder trial, 245–50, 252, 271
photo evidence for, 212–19, 234, 243–44, 268

Kaiser, John Boynton
and Arbuckle murder investigation, 100, 101, 124–25
career of, 38–39
and death of Heinrich, 274
Heinrich's correspondence with, 29, 53, 81–83, 219
and Heinrich's financial struggles, 41
and Heinrich's morale, 236
and Heinrich's position as college instructor, 202
Heinrich's relationship with, 38–39, 40–41, 83
and Heinrich's stiff presentation style, 236–37
and Heinrich's writings, 153–54
and Heslin murder case, 85
and Hindu Ghadar Conspiracy, 39–40
intellectualism of, 101
and Lamson investigation, 21
marriage of, 82
reference works provided to Heinrich by, 39–40, 82–83, 148, 188, 230
and stamp collecting, 40
Kelley, Sara, 238, 249
Keza, Josephine, 108
Kirk, Paul, 263–64, 267
Kneeshaw, R. Stanley, 243
Kyka, Thomas, 37, 38

laboratory of Heinrich. *See* Heinrich Technical Laboratories
Lamson, Allene Genevieve "Bebe"
custody battle over, 254, 255, 258
at father's arraignment, 227–28
and father's arrest, 19

Lamson, Allene Genevieve "Bebe" *(cont.)*
parents' fates concealed from, 258
reunited with father, 261–62
at sleepover with grandmother, 11, 12
Lamson, Allene Thorpe
background of, 9–10
death of, 11–17, 233–35, 265
former love life of, 239–40
journal used as evidence, 238, 239
lifestyle of, 10–11
marriage of, 19
Lamson, David
arraignment of, 227–28
and bonfire, 7, 9
and book about death row, 260–61, 262, 263
conviction and sentence of, 250, 253
and death of wife, 11–17
defense team of, 19, 243, 252–53, 254, 256, 258
in jail/prison, 17, 18–20, 255, 258–61, 263
and Lamson Defense Committee, 256
lifestyle of, 10–11
marriage of, 19
and philanderer accusations, 237–38, 264
police interrogation of, 16
post-release life of, 262–63
public supporters of, 255–56
released from prison, 261
retrials of, 259–60
reunited with daughter, 261–62
and shooting accident in youth, 239
trial of, 237–45
verdict appealed to State Supreme Court, 254, 256–58
See also Lamson murder case
Lamson, Jennie, 253–54, 255, 258
Lamson, Margaret, 228, 254, 262
Lamson murder case
alternative theory, 265
and arraignment of Lamson, 227–28
blood pattern analysis in, 20–21, 22, 230–33, 244–45, 249–50
conviction and sentencing, 250, 253
and crime scene management, 17, 21
death of Allene, 11–17
and denial of retrial, 252–53
Heinrich's investigation of, 20–21, 22, 23, 229–36
Heinrich's reaction to verdict, 250, 254
Heinrich's reenactments for, 20, 244, 254, 257–58, 259
Heinrich's trial testimony in, 243–45, 257–58, 259
innocence of Lamson defended by Heinrich, 233, 235
joint forensic tests conducted in, 230
and jury deliberations, 245–50, 252, 271
media coverage of, 17–18, 235–36, 237–38, 239–40, 249–50
and motives proposed by the prosecution, 237–39, 249
and pipe used as weapon, 237, 240, 241–42
police investigation of, 16–17
and Proescher's contradictory testimony, 240–42
retrials of, 257, 259–60, 261
rumors surrounding, 21
and Sheppard case comparisons, 264
trial of, 237–45
verdict appealed to State Supreme Court, 254, 256–58
Landini, Silvio, 62
Larson, John, 73–75, 76–77
law enforcement
archaic methods of, in 1920s, 4–5
and lack of resources for state and local, 269–70
and reforms efforts of Vollmer, 34–35
and school for police, 35–37
turn-of-the-century standards in, 34
Lazarus, Sylvain, 109
Lee, C. C. (dentist), 204–5
Lee, Clarence (detective)
and Ferguson's dismembered ear case, 188
and Schwartz murder investigation, 179–80, 182–83, 184
Leland, Thomas, 104
Lindauer, Arthur, 214, 216–17
Lindbergh, Charles, 18, 168–69

Lindsay, Allan, 237, 239, 242, 245, 251–52
Locard, Edmond, 70, 114
Locard's Exchange Principle, 70
Lynn, George, 58–63, 65, 78

Mason, Dolly, 59, 60, 65, 67
Matheson, Duncan, 93–94
Matlock, Joe, 247
May, Luke, 223–24
McCarthy, John, 207–8. *See also* Colwell murder case
McGovern, Chauncey
  and Arbuckle murder investigation, 112
  and Colwell murder case, 214, 223
  death of, 229
  Heinrich's collection of articles on, 155
  and Heslin murder case, 53–55, 57
  as rival of Heinrich, 82–83, 214–15, 223
McNab, Gavin
  and Arbuckle's testimony, 116
  and Heinrich's testimony, 112–14
  and Rappe's past, 111
  and retrial of Arbuckle, 123, 125
mental illness, 70–71
Michels, Lara, 1, 2
microscopes of Heinrich
  comparison microscopes for ballistics, 210–11, 212, 215–16, 218, 268
  in courtroom settings, 217–19
  producing photographic evidence with, 212, 215–17, 218, 268
  variety of, in laboratory, 3
movies, 100

National Academy of Sciences (NAS)
  on ballistics, 218
  on blood pattern analysis, 265
  on handwriting analysis, 52
  recommendations for forensic sciences/labs, 269, 270
Newton Boys, 47
Northwestern University, 223

O'Brien, Daniel, 61, 63, 93
obsessive-compulsive personalities, 2, 29
O'Connell, Dan, 151, 157
organized crime, 133, 223
Osborn, Albert, 202–3

Peterson, George, 246, 247, 248
petrographic analysis procedure, 67–70, 80–81, 189, 190–92
Philosophy of Eternal Brotherhood, The, 180
Place, Julia, 14–15
poisonings, 208
polygraph tests, 72–78
Popp, Georg, 69
poroscopy, 114
post-mortem index (PMI), 188–89
Prevon, Zey
  and Arbuckle's testimony, 116
  and Bender, 95
  and district attorney's case against Arbuckle, 98–99
  grand jury testimony of, 104–5, 106–8
  placed in protective custody, 105
  police interview of, 91–93
  trial testimony of, 111–12
Proescher, Frederick, 230–31, 240–42, 249, 250
profiling
  and Ferguson murder case, 197–98
  and Heslin murder case, 59–60
  and Hindu Ghadar Conspiracy, 40
  history of, 56
Prohibition
  availability of alcohol during, 47, 60
  and crime in the 1920s, 4, 47
  and organized crime, 133
  repeal of, 223
  and social mores, 118, 169
  and Wickersham Report's condemnation of, 223

Rappe, Virginia
  and Arbuckle's defense strategy, 111
  and cause of death, 103, 106
  death of, 87–88, 89–90
  media coverage of, 91
  police investigation of death, 90–91, 92–93
  See also Arbuckle murder case

Rea, Edwin, 241, 243, 245, 251–52
rigor mortis
  and Lamson investigation, 16, 234
  and Schwartz investigation, 172, 178
Roberts, Mary Dolores, 237
Rowe, Gordon, 199–200, 201
Ruedy, Alfred H., 173

Sacco and Vanzetti case, 218
Saier, Milton, 234
Salvation Army, 119–20, 122
sand, petrographic analysis of, 67–70, 80–81, 189, 190–92
San Francisco, California, 43
San Francisco Examiner, 58, 59–60, 65, 190
San Quentin Prison, 255, 256, 258, 260, 263
Schwartz, Alice Orchard Warden
  and affair of husband, 182
  identification of body, 173, 177
  marriage of, 166
  and suicide of husband, 183
Schwartz, Charles Henry
  background of, 166, 182
  faked death of, 168, 172
  and funding from investors, 169
  and heart-balm lawsuit following affair, 164–65, 166–67, 182
  and history of chats with crime investigators, 179–80, 184
  identified as murderer, 181
  manhunt for, 182
  marriage of, 166
  police investigation of, 182
  suicide of, 182–83
  and synthetic silk formula, 163–64, 167–68
Schwartz murder case
  blood analysis in, 176, 180
  and faked death of Schwartz, 168, 172
  financial motive for, 181
  and fingerprints of victim, 172, 178
  and hair analysis, 178
  Heinrich's investigation of, 172–79, 180–81
  identification of body, 173, 177–79, 181
  Ponzi scheme exposed, 175
  suspect list in, 173, 174, 175
Scotland Yard, 56, 197
Semnacker, Al, 107, 108
Seng, Marvin, 139, 142
sexual assaults, 100, 118. *See also* Arbuckle murder case
sexual norms, 118
Sharpe, Dick, 239
Sheppard, Sam, 264
Sherlock Jr. (film), 128
Sherman, Lowell, 90
Shirley, Doris (later Putnam), 73, 79–80
Simplified Blood Chemistry as Practiced with the Ettman Blood Chemistry Set, 230
Siskiyou train robbery case
  attempted heist, 138–43
  capture of DeAutremont brothers, 160–61
  dynamite explosion, 140–41, 142
  evidence left behind, 143, 144, 146
  and hair analysis, 148, 152, 158
  Heinrich's deductive reasoning in, 268
  Heinrich's investigation of, 144, 147–52, 157–59
  Heinrich's notoriety gained from, 159, 160, 161
  and history of train robberies, 132–33
  manhunt following attempted heist, 143, 145–46, 158–59
  media coverage of, 146, 159, 160
  and murders of eyewitnesses, 140, 141, 142, 143
  police investigation of, 146–47
  See also DeAutremont, Hugh; DeAutremont, Ray; DeAutremont, Roy
stamp collecting, 40
St. Valentine's Day Massacre in Chicago, 223
Swart, Franklin, 79
Syer, Robert, 244, 253, 258, 261

Taylor, William Desmond, 145
Thoits, Hazel, 228

Index

Thomas, Roger, 185
Thompson, Gilbert, 114
Thorpe, Frank, 254, 255
time of death, determining
  and forensic entomology, 188–89, 268
  and rigor mortis, 234
toxicology, 208
trace evidence analysis
  and Hightower case, 70
  and Siskiyou train robbery case, 149, 268
train heist. *See* Siskiyou train robbery case
tuberculosis, 45–46

universities, forensic labs in, 223–24
University of California at Berkeley
  and archive from Heinrich's lab, 1–3
  Heinrich's role as instructor at, 36–37, 38, 201, 268
  Heinrich's studies at, 31
  and Kirk, 264
  School for Police at, 35–37
  School of Criminology established at, 267
University of Chicago, 224
U'Ren, Milton, 92, 93, 95, 96
US Department of Justice, 269
US Engineers' Reserve Corps, 40

Veale, Richard, 187–88
Vincent, Helen, 14–15
violent crimes, in 1920s, 4–5
Vollmer, August
  and Arbuckle murder investigation, 97, 99, 123
  background of, 37
  faculty position at University of Chicago, 224
  as "father of modern policing," 34, 224
  and Ferguson murder case, 197
  and heart-balm lawsuit, 165–66
  Heinrich's relationship with, 35–36, 37, 38
  and Hollywood film industry, 100
  impact on criminal justice field of, 34–35, 267
  and Kirk, 263–64
  and Lamson case, 21, 254, 256
  and polygraph test of Hightower, 73, 74, 75, 77
  reforms led by, 34–35
  and school for police, 35–37
  and Schwartz murder investigation, 179
  strong skills of detectives under, 188
  and Wickersham Report on Prohibition, 223

Waite, Charles, 210
Waste, William H., 257
Weber, George A., 20, 22, 244
Weber, Jean, 20
Wendel, Marie
  and arrival of Heslin's murderer, 45
  and departure of Heslin, 46
  housekeeping duties of, 44
  and identification of Hightower, 75
  and media coverage of case, 50
  reporting Heslin's failure to return, 47
  trial testimony of, 78–79
We Who Are About to Die: Prison as Seen by a Condemned Man (Lamson), 260–61, 262, 263
Whirlpool (Lamson), 262
*Why I Want to Trave*l (Heinrich), 153
Wilkens, Anna, 144–45
Willingham, Cameron Todd, 270
women, Heinrich's perspectives on, 118
Woollcott, Alexander, 260–61
World War I, 39–40, 46
Wright, Ralph Wesley, 11
writing ambitions of Heinrich
  detective stories, 152–54, 159–60, 193
  juvenile fiction, 171
  plays, 153
  poetry, 153
  and publications in forensic science field, 171, 273
wrongful convictions, 269, 270, 271

Zink, Howard, 16